FLOWER POLLINATION
IN THE
PHLOX FAMILY

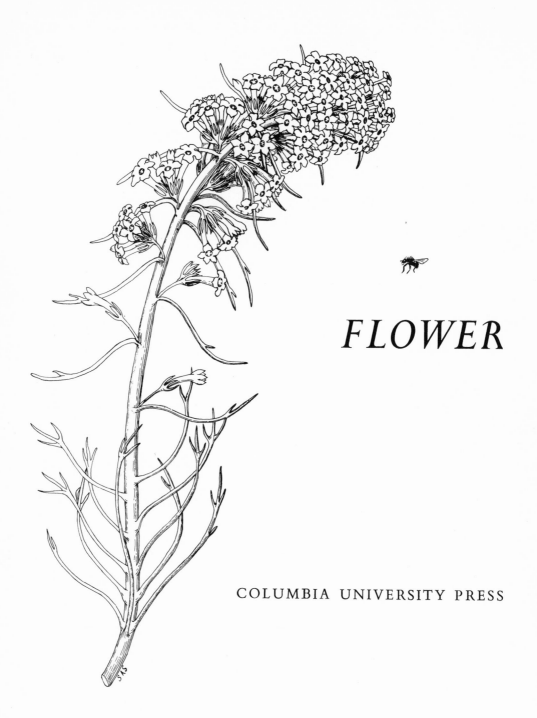

FLOWER

COLUMBIA UNIVERSITY PRESS

POLLINATION
IN THE PHLOX FAMILY

VERNE GRANT AND

KAREN A. GRANT

NEW YORK AND LONDON 1965

Verne Grant is Geneticist at Rancho Santa Ana Botanic Garden, Claremont, California. Karen A. Grant is a research biologist.

Figures 6, 23, 25, 27, and 29 are reprinted from *Illustrated Flora of the Pacific States*, vol. iii, by LeRoy Abrams, with the permission of the publishers, Stanford University Press, copyright 1951 by the Board of Trustees of Leland Stanford Junior University.

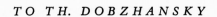

TO TH. DOBZHANSKY

PREFACE

We thus see that the structure of the flowers of Orchids and that of the insects which habitually visit them, are correlated in an interesting manner. CHARLES DARWIN, 1862

This research monograph deals with the relationships between flowers and animals, and with some consequences of those relationships for flower evolution, in terms of a single representative plant family which has been extensively studied.

In Chapters 2 through 9 we describe how 122 of the 327 species in the Phlox family are pollinated. The factual basis for these descriptions has been provided primarily by a recent program of floral ecological research, yielding pollination records for 109 species, and secondarily by some valuable previously published studies of other workers on 13 additional species. The type of breeding system has been determined experimentally for 93 of these species and is likewise recorded in Chapters 2 through 9. A brief overall view of the modes of cross-pollination and self-pollination in the family is given in Chapter 10.

In Chapter 11 we relate the floral ecological findings to the known natural affinities and probable phylogenetic trends in the Phlox family. This exercise enables us to consider the probable course of evolutionary changes in the method of pollination. With these considerations as a basis, we then go on to discuss some of the controlling factors involved in the evolution of pollination systems.

Flower Pollination in the Phlox Family is related to, though independent of, a previous monograph, *Natural History of the Phlox Family: Systematic Botany*, written by one of us (V. G.) and published in 1959. The latter work presents a generalized system of classification of the Polemoniaceae, and discusses probable phylogenetic trends and migrational history in the family. The present study builds on the taxonomic foundations laid in the earlier volume. Some readers of *Flower Pollination in the Phlox Family* may therefore wish to refer to the companion work for taxonomic details and background which fall outside the scope of the present monograph.

VERNE GRANT

Claremont, California
KAREN A. GRANT

August, 1964

ACKNOWLEDGMENTS

This investigation was carried out at the Rancho Santa Ana Botanic Garden as the home institution and base of operations for field trips.

The work was supported from 1959 to 1965 by research grant 9962 from the National Science Foundation. This grant met the costs of the field work, artwork, and manuscript preparation. The research and the publication of the results would not have been possible without the financial support of the National Science Foundation and the helpful personal interest of its officials, particularly Dr. Walter H. Hodge and his staff in the systematic biology section.

Our very special thanks go to the following men who identified insects in various groups: Lloyd Martin (Lepidoptera); P. H. Timberlake (bees); Charles Papp (beetles); Jack Hall (Bombyliidae); Curtis W. Sabrosky (scavenger flies); H. E. Milliron (bumblebees); David P. Gregory (bees).

We also wish to acknowledge the fine work of the three biological artists who made the drawings for this book. Charles Papp of Riverside, California, made the insect drawings. Richard Beasley prepared the color plates and drew the hummingbirds and many of the plants. And Shirley Schuyler made many of the plant drawings.

We enjoyed the use of the field station of the American Museum of Natural History in the Chiricahua Mountains, Arizona, during one phase of the work.

Numerous botanists have contributed valuable unpublished observations, as noted in various places throughout the text.

We wish to acknowledge, finally, the helpfulness and efficiency of the editor and publisher.

CONTENTS

I

INTRODUCTION

THE PROBLEM

Many families, tribes, and genera of angiosperms contain a rich diversity of flower forms. This diversity has been regarded in different lights by different botanists. It has traditionally been taken for granted in plant taxonomy as a useful guide in working out the system of classification; it has been viewed as a source of materials for ornamental horticulture; it has been regarded in some quarters as a product of alleged orthogenetic tendencies in the plants. The evolutionary plant biologist is more apt to see the floral diversity as the result of natural selection taking a different course in different phyletic lines.

The selective factor which is likely to play the principal role in the molding of flower forms is the agent of pollination. The diversification of related plants with respect to their floral characters can be regarded in this view as a process of adaptive radiation in the method of pollination.

This thesis would be greatly strengthened by the finding of a general correspondence between particular types of floral mechanisms and particular kinds of pollinating agents within a heterogeneous natural plant group. Now a group larger than a medium-small family will prove difficult to survey thoroughly; and a group as small as a genus will not teach us much about macroevolution; but a group the size of a tribe or medium-small family can be handled as a unit and may yield results of general significance. The question then is this: Do the various flower forms within a diversified family or tribe represent floral mechanisms specialized for pollination by different kinds of agents?

The data necessary to answer this question have heretofore never been assembled on an adequate scale for any plant family or tribe. Darwin's pioneering work on the orchids (1862) dealt of necessity with a spotty sampling of a very large family. The surveys of pollinating mechanisms in the Scrophulariaceae (Pennell, 1935), Cactaceae (Porsch, 1939), and Orchidaceae-Ophrydeae (Vogel,

1959), while extremely interesting and suggestive, are based primarily on floral morphology, and not to a sufficient extent on actual pollination records. A recent survey of floral evolution in the Ranunculaceae (Leppik, 1964), as influenced by pollinating agents, is likewise most interesting and suggestive; but here again the observational evidence presented is insufficient to support the conclusions reached. Müller's (1881) studies of various families in the Alps set a model for field observations but were restricted to a single small geographical area.

The Polemoniaceae is large enough to include numerous and diverse forms, with about 327 species in 18 genera, and to have colonized major areas on three continents, yet it is also small enough in size to be susceptible to treatment as a unit. Furthermore, the phylogenetic relationships in at least some branches of the family can be inferred with reasonable confidence (see Grant, 1959). For a considerable number of years we have been systematically collecting information on the types of floral mechanisms, breeding systems, and modes of pollination in this family (see Grant, 1961, for a preliminary account). The known pollination systems can thus be related to the probable phylogenetic affinities. This correlation, we believe, enables us to describe a large part of the evolution of the Phlox family as a series of specializations for different methods of pollination.

GENERAL FLORAL CHARACTERS

The basic floral plan in the Polemoniaceae is five sepals, five petals, five stamens, and a superior pistil composed of three carpels. The sepals are united into a tubular calyx and the petals into a tubular corolla. The corolla is 5-lobed and radially symmetrical in nearly all species. The stamens are fused at the base to the corolla wall. The carpels are united into a 3-celled ovary, a single style, and a 3-lobed stigma.

Nectar is produced by a glandular disk surrounding the base of the ovary and collects in the base of the corolla tube.

The anthers and stigma of the same flower ripen in protandrous order in most cases. The protandry may be complete or partial with a short period of overlap between the male and female stages. Or, in many autogamous species, the anthers and stigma mature simultaneously or essentially so.

The flowers are usually grouped in a cymose inflorescence, which may be loose or congested.

These are the common floral characteristics of nearly all Polemoniaceae. On this basic theme, however, evolution has composed the numerous variations which we consider in later chapters.

SYSTEM OF CLASSIFICATION

It may be useful for purposes of reference to present here the system of classification adopted in *Natural History of the Phlox Family* (Grant, 1959). The recognized genera and sections are listed together with the particular species discussed in this volume.

I. Cobaea Tribe

1. Cobaea.
   (a) Cobaea. *C. scandens; trianaei; lutea.*
   (b) Aschersoniophila.
   (c) Rosenbergia. *C. penduliflora.*

II. Cantua Tribe

2. Cantua. *C. quercifolia; pyrifolia; candelilla; buxifolia.*
3. Huthia. *H. longiflora; coerulea.*

III. Bonplandia Tribe

4. Bonplandia. *B. geminiflora.*
5. Loeselia.
   (a) Loeselia. *L. glandulosa; coerulea; mexicana.*
   (b) Glumiselia. *L. grandiflora.*

IV. Polemonium Tribe

6. Polemonium.
   (a) Polemonium. *P. reptans; caeruleum; foliosissimum; boreale; delicatum; californicum; pulcherrimum; mexicanum; pauciflorum.*
   (b) Melliosma. *P. viscosum; confertum; eximium; elegans.*
   (c) Polemoniastrum. *P. micranthum.*
7. Collomia.
   (a) Collomiastrum. *C. mazama.*
   (b) Collomia. *C. grandiflora; cavanillesii; linearis.*
   (c) Courtoisia. *C. heterophylla.*

8. Allophyllum. *A. glutinosum; divaricatum; integrifolium; gilioides; violaceum.*
9. Gymnosteris. *G. parvula; nudicaulis.*
10. Phlox.
    (a) Phlox. *P. glaberrima; paniculata; maculata; stolonifera; subulata; stansburyi; dolichantha.*
    (b) Divaricatae. *P. divaricata; pilosa; nana; drummondii; cuspidata; roemeriana.*
    (c) Occidentales. *P. caespitosa; diffusa; multiflora; andicola.*
11. Microsteris. *M. gracilis.*

V. Gilia Tribe

12. Gilia.
    (a) Giliastrum. *G. rigidula; foetida; incisa; stewartii.*
    (b) Giliandra. *G. pinnatifida; pentstemonoides; leptomeria; hutchinsifolia; micromeria.*
    (c) Gilia. *G. tricolor; angelensis; achilleaefolia; capitata; clivorum; millefoliata; nevinii; laciniata; valdiviensis.*
    (d) Arachnion. *G. ochroleuca; exilis; cana; diegensis; brecciarum; latiflora; tenuiflora; leptantha; aliquanta; austrooccidentalis; clokeyi; crassifolia; inconspicua; interior; jacens; malior; mexicana; minor; modocensis; ophthalmoides; sinuata; tetrabreccia; transmontana; tweedyi; flavocincta.*
    (e) Saltugilia. *G. caruifolia; splendens; leptalea; australis; stellata; capillaris.*
13. Ipomopsis.
    (a) Phloganthea. *I. multiflora; tenuifolia.*
    (b) Ipomopsis. *I. rubra; arizonica; aggregata; tenuituba; candida; thurberi; longiflora; macombii.*
    (c) Microgilia. *I. congesta; spicata; pumila; polycladon; depressa; minutiflora; gossypifera.*
14. Eriastrum. *E. densifolium; sapphirinum; eremicum; luteum.*
15. Langloisia. *L. punctata; matthewsii.*
16. Navarretia.
    (a) Aegochloa. *N. atractyloides; hamata; squarrosa.*
    (b) Masonia. *N. peninsularis; viscidula.*
    (c) Mitracarpium. *N. pubescens.*
    (d) Navarretia. *N. involucrata; subuligera.*
17. Leptodactylon. *L. californicum.*

18. Linanthus.
    (a) Siphonella. *L. nuttallii.*
    (b) Pacificus. *L. grandiflorus.*
    (c) Leptosiphon. *L. parviflorus; androsaceus; bicolor; breviculus.*
    (d) Dactylophyllum. *L. liniflorus; pygmaeus; pusillus; harknessii; septentrionalis.*
    (e) Dianthoides. *L. dianthiflorus; parryae.*
    (f) Linanthus. *L. dichotomus.*

METHODS

The activity of pollinating agents and the process of pollination often occur mainly during a few relatively brief periods during the day and during the flowering season when environmental conditions are favorable and the pollinators are on the wing in numbers. The optimum periods of pollination cannot be controlled by the floral ecologist or even predicted exactly but must be discovered by observations at well-spaced intervals. In order to understand the pollination ecology of a species it is necessary to live with its populations and observe the plants at different times of day and night, under different weather conditions, and at different stages of the flowering season. The requirements for field observations, difficult as these sometimes are to obtain, were well understood by Sprengel and Müller but have been ignored by another school of workers whose publications have a rather high ratio of predictions to facts.

In the present investigation a large amount of time has been spent in the field. A number of plant species have been observed fairly extensively in the wild without obtaining adequate or even any pollination records. For the most part these cases are not mentioned in the following account. On the other hand, we have managed by sheer luck to be in the right place at the right time to see the course of pollination in full swing in some other species. As regards most of the species to be described later, a reasonable amount of observation has yielded a fairly adequate picture of the mode of pollination. The observations of other workers as cited in the text—some published, others unpublished, and often from geographical areas which we have been unable to visit—have supplemented our own observations in a most valuable way.

The floral mechanisms have been studied in the natural plant populations and wherever possible again in garden-grown plants. The descriptions and measurements given in the text refer to the live flowers of particular populations, which may or may not be typical of the species as a whole.

The proboscis lengths of insects given in the text are likewise usually those of specimens associated with the particular species and population of Polemoniaceae under discussion, and do not necessarily portray the range of variation in the insect species.

The problem of insect identification has been met by relying heavily on the help, generously given, of specialists in the various insect groups. This is not to say that the identification problem has been solved completely. Entomologists are busy with projects and official duties of their own; identification in many groups of insects is a difficult and time-consuming matter; and in some groups the taxonomy has not been worked out yet to a point where identification is possible or meaningful. For all these reasons the insects have frequently been identified to genus, or in some cases in the Diptera to family, rather than to species. There are cases, moreover, like *Bombylius lancifer*, where a species name can be used, but applies to what appears to be an unanalyzed complex of forms. The insect specimens obtained during this investigation have either been deposited in the collections of specialists or been retained in the working collection of the authors where they are available for further study.

It has been necessary to follow a single system of classification of insects consistently in this work. Different authorities often disagree as to whether a group of bees or of butterflies should be treated as a genus or subgenus. On such questions we of course have no opinion of our own, but have simply followed some standard reference work. In nomenclature also we have followed standard checklists such as Muesebeck, Krombein, and Townes (1951) for bees and McDunnough (1938) for Lepidoptera.

As a general rule we have not collected or listed predatory beetles, bugs, spiders, and the like which are found on many flowers but do not feed on them directly or carry pollen.

For as many species as possible the plants have been grown in the experimental garden and artificially self-pollinated to determine whether they are self-incompatible or self-compatible and, in the latter case, further tested for the presence or absence of autogamy.

The information on floral mechanisms, animal visitors, and breeding systems was compiled on file cards for each species and race of Polemoniaceae. The facts presented in the following chapters have been taken directly from the cards with many unnecessary or irrelevant details omitted.

In the presentation of the results in Chapters 2 through 9, the details are given in reduced type under each plant species. There is a briefer and more general summary in text type at the beginning of the account of each species, which may be sufficient in itself for the purposes of many readers. A still more general summary and overall view of the pollinating mechanisms in the Phlox family is found in Chapter 10. Chapter 11, finally, is devoted to an evolutionary interpretation of the findings.

2

POLEMONIUM

Polemonium is a genus of about 23 species, most of which are perennial herbs. The flowers in most species are blue, campanulate or funnelform with a broad throat, have exserted stamens and styles, and provide nectar concealed under a tuft of hair in the short corolla tube. Pollination by bumblebees and other bees is the basic mode, though other pollination systems have developed within the genus.

SECTION POLEMONIUM

Polemonium is divided into three sections. The first of these, the section Polemonium, consists mainly of moderately tall perennial herbs with broad pinnate leaves. These plants typically grow in meadows, bogs, or moist ground in temperate regions.

Polemonium reptans   *Fig.* 1

This small perennial herb is common in meadows and moist woods of the eastern United States from the Mississippi valley to the Atlantic coast, where it blooms in spring from April to May or June. The light blue, campanulate flowers are borne in a loose cyme and lie canted to the side. The plants are self-incompatible.

Robertson (1891, 1928) made extensive observations of flower-visiting insects in southern Illinois. The flowers attract a wide variety of bees (37 species are recorded), as well as miscellaneous Diptera and Lepidoptera (5 species in each order), and a few beetles. Most of these insects are regular visitors to the flowers of *Polemonium reptans*, as indicated by their presence repeatedly in different years.

The beetles and some of the flies feed on the pollen but do not regularly pollinate the flowers. The butterflies and moths similarly exploit the flowers for nectar without regularly transferring pollen.

Fig. 1. Polemonium reptans and Bombus americanorum (Apidae). Life size; flower at right × 2.

The bees and several flies, on the other hand, whether sucking nectar or collecting pollen, settle down in the flowers in such a way as to brush the anthers with their bodies and pick up pollen. The native visitors which are at once most frequent and most effective in pollination are various Apidae of the genera Bombus, Ceratina, Nomada, and Tetralonia. One megachilid bee (Osmia) and several short-tongued bees (Andrena, Augochlora) are also frequent and effective visitors. The pollination of *Polemonium reptans* is thus carried out chiefly by bumblebees and several other kinds of bees, with supplementary help from a host of other bees and a few dipterans.

FLORAL MECHANISM. The campanulate, light blue corolla is about 1 cm. long and 1 cm. broad in the limb. The slightly unequal stamens stand out just beyond the entrance to the corolla throat, the uppermost stamens being shorter than the lower ones. The stigma is exserted a short distance beyond the anthers. The flowers are protandrous. A tuft of hairs at the base of each filament in the bottom of the corolla covers the nectar chamber, which is in the lower corolla tube, and purplish lines in the lower throat apparently function as nectar guides.

BREEDING SYSTEM. We self-pollinated and cross-pollinated 80 flowers on 2 individual plants grown in Claremont. The 40 self-pollinated flowers produced no capsules, whereas the 40 cross-pollinated flowers set 21 capsules and numerous seeds, showing that the plants are self-incompatible.

INSECT VISITORS. Observations of Robertson (1891, 1928) around Carlinville, Illinois. The Hymenoptera and Lepidoptera reported by Robertson are listed below under their modern names. The insects were mostly sucking nectar and/or collecting pollen, and were effecting pollination unless noted otherwise. Insect species collected repeatedly in different years are annotated as repeats.

HYMENOPTERA

Apidae: *Nomada affabilis. Nomada hydrophylli* (frequent). *Nomada ovata. Tetralonia belfragei* (abundant, repeat). *Tetralonia dilecta* (frequent, repeat). *Anthophora ursina. Ceratina calcarata* (repeat). *Ceratina dupla* (abundant, repeat). *Bombus americanorum* (abundant, repeat). *Bombus auricomus* (repeat). *Bombus griseocollis* (frequent, repeat). *Bombus impatiens* (frequent, repeat). *Bombus ridingsii* (repeat). *Bombus vagans. Psithyrus variabilis* (repeat). *Apis mellifera* (abundant, repeat).

Megachilidae: *Robertsonella simplex. Osmia atriventris* (repeat). *Osmia conjuncta* (repeat). *Osmia lignaria* (repeat). *Osmia pumila* (abundant, repeat).

Halictidae: *Agapostemon radiatus* (repeat). *Augochlora pura* (frequent, repeat). *Augochlora similis* (repeat). *Augochlora striata* (abundant, repeat). *Halictus rubicundus. Lasioglossum coreopsis* (repeat). *Lasioglossum coriaceum* (repeat). *Lasioglossum obscurum. Lasioglossum pilosum* (repeat). *Lasioglossum versatum* (repeat).

Andrenidae: *Andrena carlini. Andrena geranii* (frequent). *Andrena nasonii. Andrena polemonii* (abundant, repeat). *Andrena pruni* (repeat). *Andrena sayi.*

DIPTERA

Syrphidae: *Mesogramma marginata* (feeding on pollen, not pollinating). *Pipiza femoralis* (feeding on pollen, not pollinating, repeat). *Rhingia nasica* (feeding on nectar and pollen).

Empididae: *Pachymeria pudica.*

Bombyliidae: *Bombylius major* (repeat).

LEPIDOPTERA

Pieridae: *Colias philodice* (frequent, not pollinating, repeat).

Hesperiidae: *Erynnis brizo* (not pollinating). *Erynnis juvenalis* (frequent, not pollinating, repeat).

Sphingidae: *Hemaris thysbe* (not pollinating, repeat).

Phalaenidae: *Autographa falcifera simplex* (frequent, not pollinating, repeat).

COLEOPTERA

Anthicidae: *Corphyra terminalis* (feeding on pollen but not pollinating).

Coccinellidae: *Megilla maculata* (feeding on pollen but not pollinating).

Polemonium caeruleum   Plate I B

Polemonium caeruleum is a perennial herb of moist or boggy places in cool or cold temperate latitudes throughout the northern hemisphere. The flowers are campanulate, sweet-scented, and colorful with a blue corolla and orange anthers. One part of the white base of the corolla absorbs and another part reflects ultraviolet, adding to the mixture of colors for the insects (Kugler, 1963). The flowers are protandrous. Strains in England are self-compatible (Pigott, 1958).

The flowers attract bees chiefly though not exclusively. In a meadow in the Sierra Nevada of California we saw butterflies feeding on the flowers of Aster and Achillea but passing over the Polemonium plants in their paths. Some beetles find their way into the flowers for nectar or pollen.

The usual pollinators are bumblebees and other medium-large bees. These insects settle down on the flowers, clinging partly to the essential organs and partly to the corolla, while they collect pollen or probe for the nectar concealed under a covering of hairs in the corolla tube. In the course of this action they brush the anthers or stigma with their venter. The individuals of Megachile working on Polemonium in the Sierra Nevada had large amounts of the sticky orange pollen on their venters. In settling down on a flower in the pistillate condition the pollen-coated venters of these bees normally came into contact with the upturned stigma.

FLORAL MECHANISM. The flowers are grouped in congested terminal cymes and point outward or slightly downward. They are faintly sweet-scented. In color they appear blue with a white base to us. Kugler (1963, 302) has found that the blue part reflects ultraviolet, while part of the white base reflects ultraviolet, and another part of the white base absorbs ultraviolet. Therefore the two color zones perceived by us break up into three contrasting color zones for the insects.

The corolla is campanulate and consists of a short tubular base and spreading lobes. The limb varies from 1.6 to 3 cm. in diameter in the different races.

The nectar is secreted by a green disk surrounding the base of the ovary and accumulates in the short corolla tube. The base of each stamen filament where it joins the corolla is hairy. Collectively these tufts of hairs form a protective roof over the bulk of the nectar. Some nectar spreads out by capillarity on the upper surface of the hairy zone. The purple lines converging from the base of the corolla lobes into the tube in *P. c. amydalinum* and some strains of *P. c. caeruleum* (Knuth, 1909, 110) may function as nectar guides.

The stamens are exserted to varying degrees in the different races, and the orange anthers form a color contrast with the blue of the corolla. The style is declined on the lower side of the corolla and bends upward at the end, the stigma standing out beyond the anthers. The flowers are protandrous. In *P. c. caeruleum,* according to Pigott (1958) and Müller (1881), the five anthers dehisce successively and shed pollen over a period of a day or two as the corolla opens; then when the anthers have shriveled, the style, hitherto curved down and immature, bends upward and opens its stigma.

BREEDING SYSTEM. Pigott (1958) states that *P. c. caeruleum* in England is self-compatible.

Self-pollination unassisted by insects is probably not impossible, but was not witnessed by Müller in four seasons of observation, nor by ourselves in garden-grown plants, and is definitely minimized by the protandry and the spatial separation of anthers and stigma. Ekstam (1897, 121) observed occasional short-styled flowers of this species on Novaya Zemlya in which self-pollination occurred, but the floral mechanism in normal flowers precludes autogamy here too.

INSECT VISITORS. Davidson (1950) recognizes four geographical races treated as subspecies. Pollination records are now available for all of the subspecies except the eastern American *P. c. vanbruntiae*. The European race for which data are listed below is *P. c. caeruleum,* the Alaskan race *P. c. villosum,* and the Californian race *P. c. amygdalinum.*

ENGLAND, NATURAL HABITATS   (Pigott, 1958)

Apidae: *Bombus lucorum* (most common species). *Bombus lapidarius. Bombus pratorum.*
Syrphidae: *Volucella bombylans. Eristalis sp. Syrphus sp. Platycheirus manicatus.*
Panorpidae (Mecoptera): *Panorpa communis.*
Coleoptera: various small beetles.

ENGLAND, CULTIVATED   (Pigott, 1958)

Apidae: *Bombus terrestris. Bombus hortorum. Apis mellifera.*

ALPS   (Müller, 1881, 257)

Apidae: *Bombus alticola. Bombus lapidarius. Bombus lapponicus. Bombus pratorum. Bombus terrestris. Apis mellifera.*
Megachilidae: *Megachile sp.*
Syrphidae: *Rhingia campestris. Syrphus ribesii.*
Cerambycidae (Coleoptera): *Pachyta interrogationis* (eating the anthers).

NORTHERN EUROPE (Knuth, 1909, 111)

Apidae: *Bombus hortorum. Bombus lapidarius. Bombus terrestris. Bombus pratorum. Apis mellifera.*

Megachilidae: *Chelostoma campanularum. Chelostoma nigricorne. Coelioxys sp. Osmia rufa. Megachile sp.*

Telephoridae (Cantharidae) (Coleoptera): *Dasytes flavipes.*

FAIRBANKS, ALASKA: *Bombus sp.*

SIERRA NEVADA, CALIFORNIA: *Megachile sp.*

Polemonium foliosissimum

This perennial herb of the Rocky Mts. is similar to *Polemonium caeruleum* in its general vegetative and floral characters. The flowers are campanulate, with a blue-violet corolla and yellow anthers, and are protandrous and self-compatible. The main pollinators of a population at Gothic, Colorado, are bumblebees, particularly *Bombus flavifrons* and *B. rufocinctus*, which are abundant visitors and effective pollen carriers. These bees are active on the flowers during the warm part of the day from midmorning on. They perch on the stamens to probe for nectar or gather pollen; in either case some pollen invariably adheres to their venters, which come into contact with the stigmas of other flowers in the pistillate condition.

FLORAL MECHANISM. The flowers and inflorescence are similar to those of *Polemonium caeruleum*. The flowers close at night and open in midmorning. When open they are fragrant and blue and yellow, or sometimes white, with a campanulate form. The filament bases bear tufts of hair which cover the nectar in the tube below. The stamens and style are exserted on the lower side of the corolla. The flowers are protandrous, but anthers and stigma overlap in time of ripeness, and the recurved stigma may touch the long lower anther in some flowers.

BREEDING SYSTEM. Davidson (1950) found from experiments on a strain from European botanic gardens grown in Berkeley, California, that *Polemonium foliosissimum* is self-compatible and is capable of some autogamous selfing. We have confirmed these findings with a strain from the Wasatch Mts., Utah.

INSECT VISITORS. The following insects were observed in the Rocky Mt. Biological Laboratory, Gothic, Colorado, on several successive days in July 1961. One of the bee species, *Andrena atala*, has also been found on *Polemonium foliosissimum* near Boulder, Colorado, by Dr. U. N. Lanham (personal communication).

Apidae: *Bombus flavifrons* (abundant, effective pollinator). *Bombus rufocinctus* (abundant, effective).

Megachilidae: *Hoplitis fulgida.*

Andrenidae: *Andrena atala* (collecting pollen).

Meloidae (Coleoptera): *Macrobasis unicolor* (fairly common on flowers but not pollinating much).

Polemonium boreale   *Plate I C, Fig. 2 A*

This low stocky perennial herb grows in moist rocky ground and on tundra in the arctic zone of North America and Eurasia. The large blue bell-shaped flowers are borne in clusters at the top of the stem. The plant as a whole gives off a strong musky odor, at least in some races, but the flowers have a sweet odor.

The habitats of *Polemonium boreale* are subject to long periods of cold and stormy weather when the flower-visiting insects lie low, broken by occasional short periods of fair weather and insect activity. This sporadic pattern of weather and of insect activity makes it difficult for the field biologist to work out the pollination relations of this species, and adequate field observations are not in fact available. The most common flower-visiting insects in the areas occupied by *Polemonium boreale* are bumblebees and Diptera. Both classes of insects have been seen feeding on the flowers of *P. boreale*, and both classes probably bring about pollination, but their relative efficacy as pollinators remains to be determined.

FLORAL MECHANISM. In the Alaskan plants the flowers are borne erect or ascending in clusters at the top of the stem. The large bell-shaped corolla is usually deep blue with a yellow or white center and with yellow anthers standing at the entrance. The stigma is exserted a short distance beyond the anthers.

INSECT VISITORS. In Eurasia, *Polemonium boreale* was observed on Novaya Zemlya and Spitsbergen by Ekstam (1894, 1897, 1898). Ekstam reported his observations under the name *Polemonium pulchellum*, which as presently treated does not range this far west in Eurasia, whereas the low stocky Polemonium in the area of Ekstam's observations is assigned to *P. boreale* by modern authors (cf. Davidson, 1950). In Alaska the species has been observed from the floral ecological standpoint by Dr. Ira Wiggins (personal communication) in the Brooks Range, by Dr. B. Elwood Montgomery (personal communication) on Barter Island east of Point Barrow, and by Dr. Albert Johnson and the senior author at Eagle Summit northeast of Fairbanks.

On Novaya Zemlya, Ekstam observed frequent visits by "a medium-sized fly," tentatively identified as a "Schmeissfliege" (viz., bluebottle fly or blowfly, Calliphoridae). These and other flies were the common visitors of many species of flowers in this area, including Dryas, Saxifraga, Stellaria, and Cerastium. Bumblebees were seen visiting some plants (Oxytropis, Astragalus) but never the Polemonium.

In Spitsbergen, Ekstam did not observe any insect visitors. Diptera were, however, the important pollinators of associated species of Saxifraga and Cerastium in this flora.

In Alaska, Dr. Montgomery found considerable numbers of *Bombus balteatus* visiting the flowers of *Polemonium boreale*, Dr. Wiggins saw bumblebees and an unidentified small black fly, and Dr. Johnson and the senior author did not see any insect activity.

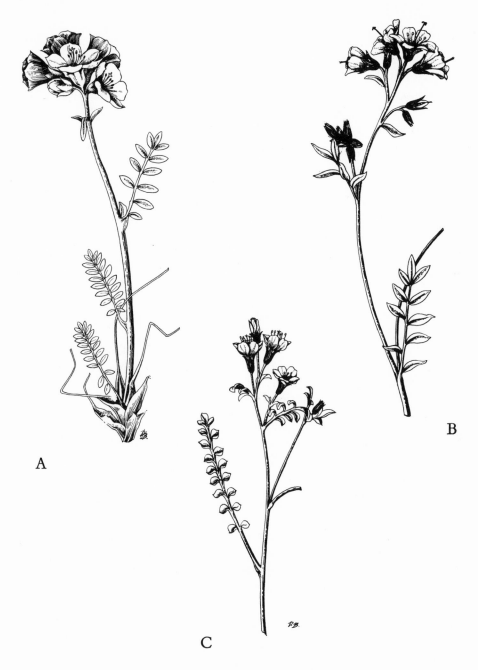

A

B

C

Fig. 2. Polemonium species. Life size.

(A) P. boreale.   (B) P. californicum.   (C) P. pulcherrimum.

Polemonium delicatum

These small perennial herbs grow in the shelter of firs and spruces in the subalpine zone of the Rocky Mts. The herbage gives off a skunk-like odor and the flowers a faint sweet scent. The corolla is pale blue with a yellow center. The flowers are open by day but close up at dusk.

At two sites in Colorado various kinds of flies have been observed visiting and pollinating the flowers. This attraction of Diptera to *Polemonium delicatum* is very consistent, and it seems possible that the skunky odor of the herbage, which is strong and easily detected from a distance, may attract the flies to the plants, whereupon they discover the flowers. A small megachilid bee was also observed as a less common visitor in one of the Polemonium populations. The flies and bees land on the flowers to feed and in so doing pick up pollen on their venters.

FLORAL MECHANISM. The herbage has a skunky odor and the flowers a faint sweet scent. The corolla is pale blue with a yellow center, campanulate, and medium-small with a limb 1.2 to 1.5 cm. broad. The stamens and style stand out in the cup formed by the limb. The flowers are open by day and close up at dusk.

INSECT VISITORS. Two populations have been observed in Colorado. One of these is in subalpine forest at an elevation of about 11,000 feet on the Buchanan Pass trail in Boulder County, and the other at the edge of alpine tundra at 11,700 feet on Trail Ridge in Larimer County.

The Buchanan Pass trail population was visited by a small green bee, *Hoplitis fulgida* (Megachilidae), and by a number of flies belonging to the Muscidae (*Thricops septentrionalis*), Anthomyiidae, Empididae, and Syrphidae. Anthomyiid and syrphid flies were observed visiting the flowers in the Trail Ridge population. These insects, both the bee and the flies, touch the stamen when they land in the flowers, and carry small amounts of pollen on their venters.

Polemonium californicum   *Fig. 2 B*

The low perennial plants grow in openings in coniferous forest at high elevations in the mountains of the Pacific slope. The medium-small campanulate flowers are similar to those of *P. delicatum*, and, as with the latter species, are visited and pollinated by various flies and bees. A strain of *P. californicum* from the Sierra Nevada is self-incompatible.

FLORAL MECHANISM. Generally similar to *P. delicatum*.

BREEDING SYSTEM. Plants in a population from Snow Flat, Yosemite National Park, California, grown and tested at Claremont, are self-incompatible.

INSECT VISITORS. Syrphid flies, dance flies (*Empis sp.*, Empididae), and an unidentified bee were observed visiting the flowers in a population on Mt. Rainier, Washington. These insects settle down on the flowers, insert their proboscis through the hairy base of the corolla into the nectar chamber below, and in the process of feeding pick up and transport pollen.

Polemonium pulcherrimum   *Fig. 2 C*

Polemonium pulcherrimum is a mat-like herb of rocky places from Alaska to the Sierra Nevada of California. In the southern part of its range it grows in the mountains just above timberline. The herbage gives off a musky scent; the flowers are sweet-scented in some races and are apparently odorless in others. The small, campanulate, blue and yellow flowers are autogamous, at least in a Sierran strain which has been tested, and produce fully vigorous inbred seedlings. In nature the flowers are also visited and cross-pollinated by bumblebees, megachilid bees, and an occasional syrphid fly.

FLORAL MECHANISM. The flowers bloom a few at a time at the outer ends of the branches, where they point upward and outward. The flowers in a population on Garfield Peak, Crater Lake, Oregon, were sweet-scented, while those near Saddlebag Lake, Mono County, California, had no detectable odor. The small campanulate corolla is blue with a yellow center. The corolla limb is about 1 cm. broad, the white anthers stand out 4 or 5 mm. beyond the corolla throat, and the style is approximately the same length as the stamens. In the Saddlebag Lake plants the stigma and anthers ripen on the same day, and self-pollination can and does occur automatically in an early stage of flowering.

BREEDING SYSTEM. The Saddlebag Lake strain was grown and tested in Claremont. The plants proved to be self-compatible and autogamous, setting seeds freely inside insect-proof breeding cages whether artificially self-pollinated or left untouched. The I_1 seedlings grown from autogamous flowers and from artificially selfed flowers were quite vigorous, and comparable in vigor to the seedling progeny of the sister crosses taken as controls. Davidson (1950, 213 and 254) mentions obtaining a different result pointing to non-autogamy in northern strains of this species.

INSECT VISITORS. In the population on the ridge above Saddlebag Lake we observed three insect visitors: *Bombus rufocinctus*, an unidentified solitary bee of medium-small size, and *Syrphus*. The Garfield Peak population was visited by *Osmia* (Megachilidae). These insects, in perching on the flowers, picked up pollen on their venters.

Polemonium mexicanum

This erect perennial or annual herb from central Mexico bears small blue and yellow flowers in terminal cymes. The anthers and stigma stand on approximately the same level just above the corolla throat. Plants grown from seeds received from English botanic gardens were self-compatible and autogamous, and the first inbred generation was vigorous. In nature the bright blue and yellow flowers

probably attract insects, which bring about some cross-pollination, and for the
rest the species probably reproduces autogamously.

Polemonium pauciflorum   Plate II C

The species of Polemonium discussed so far have blue campanulate flowers.
It is of interest to consider now a very different flower form in the genus, as
represented by *Polemonium pauciflorum* of the Sierra Madre of Mexico. This erect
herb bears trumpet-shaped yellow flowers about 3 cm. long which stand out or
hang on flexible pedicels 1 cm. or more long.

The plants are autogamous. One population on the geographical and ecological margin of the species range in Arizona is not normally visited by pollinating
animals but seems to be reproducing chiefly by autogamy. This may not be the
whole story or even the typical situation, however, since the flowers of *Polemonium
pauciflorum* have the characteristics of known hummingbird-pollinated plants, and
can be shown to be attractive to hummingbirds in artificial setups. It is probable
that the flowers are visited and cross-pollinated by hummingbirds in some parts
of the species area at least.

FLORAL MECHANISM. We have studied a cultivated strain grown from seeds obtained
from European botanic gardens and also a natural population in the Chiricahua Mts. of southeastern Arizona. The two strains are very similar but differ in some minor details.

The erect stems bear a few yellow tubular flowers which lie in an ascending or loosely horizontal
or pendant position on long flexible pedicels. In shape the corolla is a broad, gently flaring tube
3 cm. or more long, 4 to 7 mm. wide at the orifice, and 2 or 3 mm. wide in the middle. It is dull
or lemon yellow, sometimes tinged faintly with red in the limb, and brightened by golden anthers.
Kugler (1963, 303) has shown that the flowers do not reflect ultraviolet light. The nectar chamber
in the base of the tube is covered by hairs. The anthers and stigma stand at the orifice. There are
stages in the flower when the ripe stigma touches the ripe anthers automatically.

BREEDING SYSTEM. Davidson (1950, 212) has shown that the botanic garden strain is
self-compatible and autogamous. The same result has been obtained by Brand (1905) and ourselves.
The lot of I_1 seedlings raised in Claremont was vigorous. In the natural population in the Chiricahua
Mts., almost all flowers were observed to be self-pollinating and almost all flowers were setting
capsules. Since pollinating animals were absent or rare in this population, it is probably reproducing
chiefly by autogamous means.

MODE OF POLLINATION. *Polemonium pauciflorum* has all the characteristics of known
hummingbird flowers. In the one natural population observed in the Chiricahua Mts., the flowers
were not in fact visited by hummingbirds or other agents, but set seeds autogamously as we have
seen. This negative evidence is not necessarily conclusive, however, since the Chiricahua

population occurs on the geographical and ecological margin of the species area. It grows in a shady opening in Douglas fir forest where sun-loving hummingbirds do not penetrate.

When flowering branches of the Polemonium were set out in a sunny place they were quickly discovered and freely visited by several species of hummingbirds. Rufous hummingbirds (*Selasophorus rufus*) with relatively short bills 2 cm. or less long could not reach all the way down the corolla tube, but the Rivoli hummingbird (*Eugenes fulgens*) with a bill 3 cm. long fitted perfectly into the tubular flowers, and all birds in feeding contacted the anthers and stigma with the base of the bill and face.

Perhaps farther south in Mexico, where the Polemonium is more at home than in Arizona, it is visited and cross-pollinated by hummingbirds, at least to some extent.

SECTION MELLIOSMA

Tundra slopes and fell-fields above timberline in the mountains of western North America are inhabited by a series of alpine species of Polemonium treated as the section Melliosma. The plants are stocky, with densely glandular, (usually) skunky-scented herbage and small whorled leaflets, and bear (usually) blue funnelform flowers in capitate or subcapitate heads. The widespread species in the group is *P. viscosum* which ranges through the Rocky Mt. region. More narrowly distributed in the Rocky Mts. are *P. confertum* and *P. brandegei*. Similar alpine habitats in the Sierra Nevada are occupied by *P. eximium*. In the southern Cascade Mts. is *P. chartaceum* and in the northern Cascades *P. elegans*.

Polemonium viscosum

The low tufted plants grow in alpine tundra in the Rocky Mts. The herbage is not odorous but sweet-scented in this species and the flowers are very fragrant. The latter are tubular funnelform with a large throat. The corolla is deep blue-violet and the anthers orange. The flowers are visited and pollinated by various flies, chiefly Muscidae, and by bumblebees.

FLORAL MECHANISM. An extensive population growing at an elevation of 11,700 feet on Trail Ridge, Larimer County, Colorado, was studied in July 1961. The herbage is not skunky-scented but grass-like and the flowers are very fragrant. The corolla is tubular funnelform, nearly 3 cm. long, with a full throat 6 mm. broad at the orifice. The orange anthers at the entrance to the throat form a color contrast with the deep blue-violet of the corolla. The style and stigma are exserted well beyond the anthers.

INSECT VISITORS. The principal flower-visiting insects in the study area on Trail Ridge are bumblebees and flies, and we observed both classes of insects visiting and feeding on the flowers of *P. viscosum* during several days of observation. There were an unidentified large tawny bumble-bee, a smaller bumblebee, syrphid flies, and various scavenger flies. Among the latter were *Protophormia terraenovae* (Calliphoridae), *Thricops aff. villicrurus* (Muscidae), and *Pogonomyia spinitarsus* (Muscidae). These insects settle or crawl into the corolla throat, either to probe deeper for nectar or to eat or collect pollen, and leave the flowers with small amounts of Polemonium pollen adhering to their bodies.

The weather in this alpine zone is generally cold, windy, and stormy, with only occasional brief periods of warmth. The bumblebees were seen only during the noon hours. The muscid flies were active off and on throughout the day, being apparently less inhibited by cold temperatures than the bumblebees. In terms of numbers of individuals and duration of activity, the muscid flies were the most common visitors of *Polemonium viscosum*, though the bumblebees are probably more effective as pollinators when they do visit the flowers.

Polemonium confertum

This species of alpine fell-fields in the Rocky Mts. has skunky-scented herbage and bright blue and orange, fragrant flowers clustered in large globular heads. The large corollas are tubular funnelform with a cup-shaped throat and limb and a long throat and tube. The floral mechanism, particularly the spatial and temporal separation of the anthers and stigma, is such as to promote outcrossing.

A population growing in a cirque above Gothic, Colorado, was observed to be visited and cross-pollinated chiefly by a bumblebee, *Bombus rufocinctus*, and a hummingbird, probably a Calliope hummingbird, and secondarily by a mega-chilid bee, *Osmia*. The bumblebee and the hummingbird were both making regular rounds through the population every 15 to 30 minutes during the non-stormy part of the day. Although their methods of feeding are different, the hummingbird hovering on the wing and the bumblebee crawling into the throat, the proportions of the Polemonium flowers are such as to accommodate both types of visitors, and both animals were bringing about pollination.

FLORAL MECHANISM. The following observations were made in a population in a cirque basin 11,800 feet above Gothic, Gunnison County, Colorado, in July 1961.

The herbage is viscid and skunky-scented and the flowers fragrant. The flowering heads are large and globular, up to 6.5 cm. broad, and the individual flowers are blue with orange centers. The corolla is large, tubular funnelform, and has an ample throat. The limb is 2.5 to 3.0 cm. broad depending on the angle of spread of the lobes; the tube and throat together are 1.5 to 1.7 cm. long; and the lobes together with the throat form a cup. The orange anthers stand in this cup. The style lies on the lower side of the corolla and is exserted about 6 mm. beyond the anthers. Nectar is present in the lower corolla tube.

Plate I. Bee flowers. All life size.

(A) Bonplandia geminiflora.

(B) Polemonium caeruleum caeruleum.

(C) Polemonium boreale.

(D) Gilia rigidula.

(E) Gilia latiflora latiflora.

(F) Gilia leptantha pinetorum.

(G) Gilia capitata chamissonis.

(H) Eriastrum densifolium.

(I) Eriastrum luteum.

(J) Eriastrum sapphirinum.

(K) Ipomopsis multiflora.

(L) Linanthus dianthiflorus.

ANIMAL VISITORS. The plants comprising the Polemonium population about Gothic are more or less widely scattered among rocks and boulders in a small cirque basin. A medium-large bumblebee, *Bombus rufocinctus*, was making regular rounds every 15 to 30 minutes through at least part of this population, settling on and crawling into the broad corolla throat to collect food and carrying pollen on its venter. A hummingbird, probably a Calliope hummingbird (*Stellula calliope*), was also feeding on the flowers, by hovering and probing into the long and fairly broad corolla tube for nectar, in the course of which it picked up pollen on its chest. It too was making regular rounds every 15 to 30 minutes through the population and seemed to be relying on the Polemonium as one of its chief food sources, there being little else for it besides a Cirsium in the area. A green bee, Osmia, was an occasional visitor and pollinator.

Various Diptera investigated the flowers but were not seen to enter or pollinate them. Skippers and butterflies feeding on small Compositae passed by the Polemonium flowers frequently but ignored them consistently.

Polemonium eximium

This plant grows on fell-fields well above timberline in the Sierra Nevada of California. The fragrant, deep blue-violet, tubular funnelform flowers are clustered in spheroidal heads. The senior author climbed Mt. Dana several times in the summer of 1956 and 1957 in order to observe insect activity on the flowers. Syrphid flies were seen going from flower to flower with pollen on the lower thorax in both seasons, and an unidentified bumblebee was seen once.

Polemonium elegans

Polemonium elegans is a dwarf alpine perennial found on the higher peaks of the Cascade Mts. The flowers are sweet-scented and bicolored with a blue-violet corolla limb and tube and a yellow throat. On fell-fields above timberline on Mt. Rainier, Washington, in August 1957 syrphid flies were frequent and regular visitors of the flowers. The flies, in settling in the flowers to probe for nectar, became dusted with pollen on their venters.

SECTION POLEMONIASTRUM

This section contains a single species of annuals of cold desert regions.

Polemonium micranthum Fig. 3

Polemonium micranthum is a small few-flowered annual herb which has a widespread distribution in the Great Basin and in similar cold arid regions of South

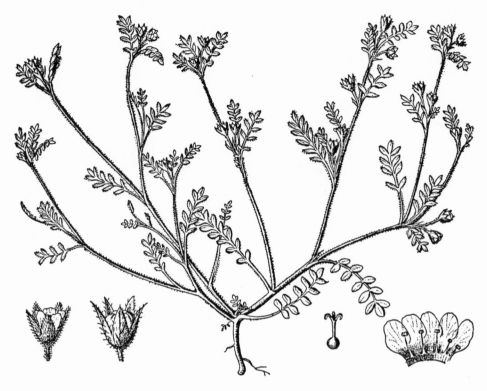

Fig. 3. Polemonium micranthum. Life size; flowers and fruit × 2; pistil and corolla × 3.
(From Brand, 1907.)

America. The small inconspicuous white flowers undergo self-pollination in the bud. The corolla later opens for a few brief hours, but this opening is biologically useless, since pollination has already occurred and the flowers are inconspicuous and not attractive to insects anyway. The plants are thus bud autogamous and highly if not exclusively inbreeding. The inbred progeny are fully vigorous.

FLORAL MECHANISM. Two strains from the Great Basin and two strains from Patagonia have been grown and studied in the screenhouse in Claremont.

The white or rarely blue corolla is inconspicuous and minute, 3 to 5 mm. long, and is included within the calyx, while the short stamens and style, in turn, are included within the corolla. In the bud stage before the corolla opens the anthers and stigma are ripe and pollen is deposited on the stigma. Self-fertilization evidently occurs in the unopened flowers. The corolla later opens in midmorning and closes again in midafternoon of the same day, never to open again, and the flower goes on to produce a fruit.

On many days during the middle of the flowering season not a single flower will be open on an individual plant. On occasional days one flower may open, or as many as three may mature simultaneously, on an individual plant.

BREEDING SYSTEM. Self-pollination occurs in the unopened flowers, and self-fertilization has probably also taken place before the corolla goes through the apparently vain motions of opening briefly. Attempts to cross-pollinate two strains artificially by emasculating the young flower buds were unsuccessful, only selfs being obtained in the progeny. The plants are thus bud autogamous and virtually completely inbreeding. The inbred progeny are fully vigorous.

3
COLLOMIA AND
RELATED GENERA

COLLOMIA

This genus containing 4 species of perennial herbs and 10 of annuals occurs mainly in western North America with outlying areas in eastern North America and temperate South America. It is related to Polemonium but more advanced than the latter in several vegetative, chromosome, and floral characters. The slender funnelform flowers are cross-pollinated by beeflies in some cases. Autogamy is also widely developed and cleistogamy is present in the genus.

Collomia mazama   *Fig. 4 C*

This endemic species of perennial herbs occurs near permanent streams in the Lodgepole pine zone on two mountains in southern Oregon. The blue-violet funnelform flowers are clustered in loose heads and give off a faint sweet odor. Outcrossing is promoted by protandry and the elongation of the mature stigma beyond the zone of anthers.

In a colony in Crater Lake National Park, Oregon, in August 1957, the flowers were being actively visited by a beefly, *Villa sinuosa jaenickiana* (Bombyliidae). The beefly settled down on the flowers and probed for nectar in the floral tube with its head and proboscis, which have an adequate reach for this task, and became dusted with pollen on its body.

Collomia grandiflora   *Fig. 4 A, B*

This annual herb occurs widely in wooded country on the Pacific slope. The stems are terminated by heads of sessile flowers subtended by leafy bracts. These

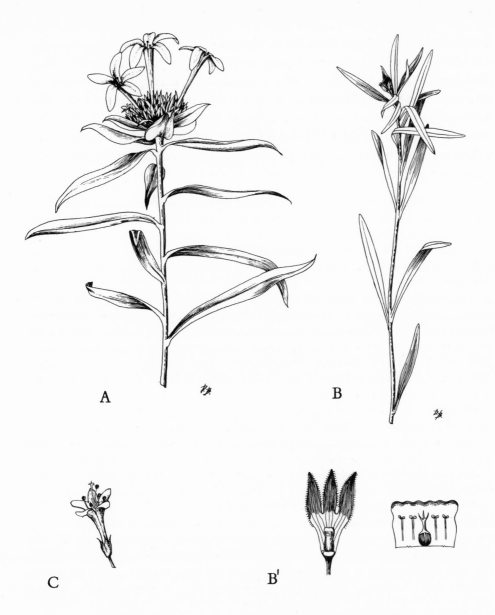

Fig. 4. Collomia. Life size; flower parts in B′ × 5 and × 7.5.

(A) C. grandiflora.   (B) Cleistogamous form of C. grandiflora.
(B′) Cleistogamous flowers enlarged.   (C) C. mazama.
                    (B′ redrawn from Ludwig, 1877.)

flowers are normally large, funnelform, and showy, with salmon-colored corollas and blue anthers. Such flowers depend upon insect visits for pollination. Tiny cleistogamous flowers are also produced occasionally, either on side branches of otherwise large-flowered individuals or on wholly cleistogamous individuals. The cleistogamous condition was first discovered by Ludwig (1877) and Scharlok (1878) in a strain from the Columbia River region naturalized in Germany, and was rediscovered by Dr. George Gillett in a natural population in the Mt. Lassen region of California.

FLORAL MECHANISM. The normal form of *Collomia grandiflora* possesses large showy flowers grouped in heads (Fig. 4 A). The funnelform corolla is salmon-colored and the anthers standing at the orifice are a contrasting blue. The corolla tube and throat are 1.5 to 2.0 cm. long and the limb is 1.0 to 1.5 cm. broad. The anthers lie at different levels in the throat on the route to the nectar, and the stigma lies close to one or more of the anthers and may come into contact with them.

A second type of flower is tiny and cleistogamous (Fig. 4 B). The colorless corolla which is about 2.5 mm. long is included within the calyx. The anthers dehisce and shed pollen directly on the stigma while the corolla is closed. The cleistogamous condition in *Collomia grandiflora* was first observed in plants introduced and naturalized in Germany by Ludwig (1877) and Scharlok (1878), and was later studied further by Ritzerow (1907; see also Hegi, 1927, 2118). The pedigree of the naturalized German strain can be traced back to seeds collected by David Douglas near the mouth of the Columbia River in Oregon.

BREEDING SYSTEM. Although the showy flowers of *Collomia grandiflora* are capable of occasional autogamous self-pollinations, plants isolated from insects and left to their own devices are not very fertile. Pollination by insects, including at least some outcrossing, is their main method of reproduction. The cleistogamous flowers, of course, produce wholly inbred progeny.

As regards the distribution of showy and cleistogamous flowers, and of their respective breeding systems, in populations, several conditions exist. Many if not most natural populations of *Collomia grandiflora* apparently consist entirely of showy-flowered plants.

Both showy and cleistogamous flowers develop on the same individual plant in some cases. The main central inflorescence then contains the showy flowers and the lateral branches the cleistogamous ones (Ludwig, 1877; Scharlok, 1878; Ritzerow, 1907). The two kinds of flowers may arise simultaneously, in which case the cleistogamous flowers develop quickly into fruits while the showy flowers are still in full bloom (Ritzerow, 1907). In plants observed by Ludwig (1877), on the other hand, the cleistogamous flowers arose early in the season and the showy flowers came out several weeks later.

Another condition observed by Ludwig is that of wholly cleistogamous individuals. These may be intermixed with outcrossing individuals or may form colonies of their own.

American botanists prior to 1957 had not recorded the occurrence of cleistogamy in native American populations of *Collomia grandiflora*. In 1957 Dr. George Gillett collected an unusual form of Collomia in Lassen National Park, California, which he later decided was cleistogamous, not knowing then of previous reports of cleistogamy in naturalized German strains. From specimens received from Dr. Gillett we could verify this conclusion. The plants look at first glance like immature *Collomia grandiflora*, but on closer inspection are seen to have sexually mature flowers in small terminal heads. The anthers and stigma mature in close contact inside the tiny corolla, and the flowers set seeds freely.

The Lassen population consists entirely of cleistogamous individuals, so far as Dr. Gillett could determine, no large-flowered plants being found in the immediate area. Normal large-flowered populations do occur in adjoining areas in Lassen National Park.

How widespread the cleistogamous Collomia is in nature remains to be seen. It is possible that botanists may have passed over such plants, taking them for immature specimens, and that the cleistogamous form is more common in nature than we can estimate from the sampling in herbaria.

Collomia cavanillesii

This annual herb of Chile and western Argentina is related to the western North American *C. grandiflora*. The flowers have a bright red limb and slender orange tube. Plants grown in Claremont from seeds sent from Chile by Dr. M. Ricardi have proved to be autogamous. If, as seems probable, the brightly colored flowers attract some kinds of insects in natural populations, this autogamy may be counterbalanced by a certain amount of outcrossing.

Collomia linearis

Collomia linearis is another annual species related to *C. grandiflora* which has a wide natural distribution throughout western North America and which in historical times has extended its range into waste places in the eastern United States and Alaska. The small, dull pink flowers are autogamous.

BREEDING SYSTEM. Comes and Peter long ago reported that *C. linearis* is self-compatible (Comes in 1879, quoted by Knuth, 1909, 114; Peter, 1891, 43). A strain from the Yosemite region of the Sierra Nevada grown in Claremont has proved to be autogamous.

Collomia heterophylla

This small branched annual occurs on the Pacific slope from Vancouver Island to Monterey County, California. The small salverform pinkish-violet flowers are autogamous but are probably also cross-pollinated to some extent by insects.

FLORAL MECHANISM. The small subsalverform corolla is pinkish-violet with darker purplish veins and often with a yellow zone in the narrow throat. The slender tube is 1.0 to 1.2 cm. long. The five anthers lie at different levels: two within the throat, two at the orifice, and one exserted slightly. The included stigma is on the same level as the two lowermost anthers and comes into contact with them automatically. The upper anthers, on the other hand, do not appear to be particularly useful for self-pollination, but may serve to deposit pollen on insect visitors and hence to promote some outcrossing.

BREEDING SYSTEM. A strain from Mt. Tamalpais, Marin County, California, grown in Claremont set full complements of seed autogamously. Some outcrossing is probably also promoted by the floral mechanism as noted above.

ALLOPHYLLUM

Allophyllum is a minor genus of annual herbs related to Collomia. Beefly pollination prevails in one species, bee pollination is present in another, and autogamy prevails in three other species.

Allophyllum glutinosum

This species occurs in the coastal mountains of southern California. Two of the races, the one inhabiting moist shaded canyons and the other growing in exposed places on bare ridges, are known to have different pollination systems.

The shaded canyon race has light blue-violet flowers with a full throat, exserted stamens and style, and slightly bilabiate form. The flowers are self-compatible but protandrous and not autogamous. They are visited and pollinated by bees and beeflies which perch on the stamen apparatus and probe into the throat and tube.

The mountaintop race has smaller flowers which do not usually attract insect visitors. This race is autogamous.

FLORAL MECHANISM. Several races differing in the details of the floral mechanism have been distinguished (cf. A. and V. Grant, 1955). Two of these, the so-called San Diego and mountaintop races, will be described here. The former occurs in wooded canyons and the latter on bare summits and ridges.

The San Diego race has the largest flowers. Its flowers are borne in pairs on slender pedicels in a loose inflorescence and are pointed outward and slightly upward. The corolla is pale blue-violet, funnelform, and slightly bilabiate as a result of the lower lobes extending forward and the upper ones reflexing backward. The stamens and style are long exserted with an upward arch. The flowers are protandrous, the style attaining its full length and the stigma opening as the stamens pass their period of pollen shedding.

The flowers of the mountaintop race are similar to the above in shape and proportions but are smaller in size and are autogamous.

BREEDING SYSTEM. A strain of the San Diego race from San Juan Canyon in the Santa Ana Mts., Orange County, California, tested in Claremont, is self-compatible but not autogamous. The mountaintop race as represented by a strain from Ortega summit in the Santa Ana Mts. is autogamous.

INSECT VISITORS. The San Juan Canyon population of the San Diego race was observed on two days in the spring of 1955. The flowers were visited by only a few insects. During the course of one morning two bees, an Anthophora and a Ceratina, and on the second morning the Ceratina and a beefly, Aphoebantus, were seen visiting the flowers. These insects alight on the exserted stamens and style which provide a suitable landing place and probe into the open corolla throat for the nectar at the base of the tube. Venter pollination results from such visits.

The Ortega summit population was also observed in 1955. A solid patch of this plant had no insect activity at all during the morning of observation. The surrounding area was covered with annuals with conspicuous flowers, such as Emmenanthe, Dicentra, Antirrhinum, Gilia, and Phacelia, and the flower-visiting insects were working on these plants but were not attracted to the inconspicuous Allophyllum. Nevertheless the latter set a full complement of seeds.

Allophyllum divaricatum   *Plate III (I), Fig. 5*

Allophyllum divaricatum is a robust annual, often with viscid skunky-scented herbage, which occurs in the foothills of the Sierra Nevada, the Coast Ranges, and the San Gabriel Mts. of California. The flowers are salverform with a slender tube varying from 7 to 9 mm. long in the San Gabriel race to 1.2 to 1.6 cm. long in the central Sierran race. The plants are autogamous.

Fig. 5. Allophyllum divaricatum (San Gabriel race) and Bombylius lancifer (Bombyliidae). Life size.

Cross-pollination is effected by various insects, chiefly *Bombylius lancifer* (Bombyliidae). This beefly inserts its straight slender proboscis, which ranges from 5 to 12 mm. long in different individuals, into the corolla tube for nectar, while hovering or lightly perching, and carries pollen on its head or tongue base.

FLORAL MECHANISM. The two races to be considered here are those of the central Sierra Nevada foothills and of the San Gabriel Mts.

In both races the small salverform flowers are borne in few-flowered clusters on the main stems and lateral branches and point upward and outward. The five anthers lie at three different levels from inside the corolla tube to above the orifice. The stigma matures at a level within the zone of anthers. Nectar is secreted and stored in the base of the slender tube.

In the Sierran race the flowers are reddish-violet with a relatively long tube, 1.2 to 1.6 cm. long, and the longest stamens stand 3 or 4 mm. above the orifice. In the San Gabriel race the flowers are blue-violet and have a shorter tube, 7 to 9 mm. long, and the longest stamens are only slightly exserted.

BREEDING SYSTEM. Strains from Groveland, Tuolumne County, in the central Sierra Nevada, and from near Claremont in the San Gabriel Range, have been tested in the Claremont screenhouse. Both strains are autogamous. The I_1 seedlings appear to be fully vigorous.

INSECT VISITORS. In a population of the Sierran race growing in an opening in yellow pine forest near Twaine Harte, Tuolumne County, in June 1954, the most common visitor was the beefly *Bombylius lancifer*. Several individuals were working the flowers systematically, passing over neighboring Clarkia and Mimulus plants to confine their operations to the Allophyllum. These beeflies hover or settle lightly on the corolla limb and their needle-like proboscis slips easily into the corolla tube. The tongues of the beeflies caught ranged from 6 to 9 mm. long. The corolla tube in this plant population was 12 mm. long, which, allowing for capillary rise of the nectar in the tube and for the insects' forcing their heads into the orifice, brings the nectar within reach of the beeflies. The head and tongue base contact the anthers and stigma.

Less common visitors and pollinators in the same population were *Lepidanthrax sp.* (Bombyliidae), feeding on nectar, and a small halictid bee, Chloralictus, collecting pollen.

Two populations of the San Gabriel race growing in the chaparral zone of this mountain range have been observed in two different years. The regular visitors are beeflies, *Bombylius lancifer*. They hover and probe for nectar, picking up pollen on the head in the process, and go systematically from flower to flower and from plant to plant.

Allophyllum integrifolium, gilioides, and violaceum

These three species of Allophyllum have small, salverform, relatively inconspicuous flowers. A representative strain of each species has been tested and found to be autogamous and to produce fully vigorous inbred progeny.

In a population of *Allophyllum violaceum* on Mt. Pinos, Ventura County, California, in June 1953, a single small Bombylius with a proboscis 4.5 mm. long was observed to visit the flowers with corolla tubes 5 or 6 mm. long, hovering and probing for nectar. Its proboscis carried pollen grains from flower to flower and left them scattered inside the corolla tube and on the stigma. Although *A. violaceum* is predominantly autogamous, therefore, it is sometimes cross-pollinated by beeflies.

GYMNOSTERIS *Fig.* 6

The minor genus Gymnosteris contains two species of diminutive annuals related to Collomia. These plants of the sagebrush plains in the Great Basin and

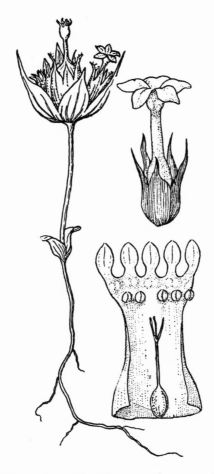

Fig. 6. Gymnosteris parvula. Plant × 5; flowers greatly enlarged.

(From Abrams, 1951.)

bordering areas have much reduced vegetative and floral characters. The small salverform yellowish flowers are borne in small heads subtended by bracts at the tips of the leafless stems.

The length of the corolla tube as given by Wherry (1944) is 1 cm. in *Gymnosteris nudicaulis* and 5 mm. in *G. parvula*. The anthers are sessile on the corolla wall at the orifice. The stigma is slightly exserted in *G. nudicaulis* and included within the tube in *G. parvula*.

Autogamy is inferred from the floral mechanism for *G. nudicaulis* and established from screenhouse tests for *G. parvula*. The inbred progeny of the latter species, as represented by a strain from Wagontire, Harney County, Oregon, are fully vigorous.

4
PHLOX

The genus Phlox consists of some 60 species of perennial herbs and subshrubs, and a few annuals, distributed widely in temperate and arctic North America. One species also occurs in Asia. This genus together with the derivative minor genus Microsteris occupies an advanced and rather isolated position within the Polemonium tribe. The closest living relative of Phlox and Microsteris seems to be Collomia, but the gap between the two phyletic lines is wide.

The distinctive features of the genus Phlox include several vegetative, seed, and chromosome characters, and also certain floral characters which are significant in relation to pollination. The corolla in Phlox is salverform with a moderately slender tube and the stamens are inserted at different levels on the corolla tube. These general characters combined with the special features peculiar to the different species fit the Phloxes for pollinating visits by various Lepidoptera.

Phlox is divided into three sections (Wherry, 1955). The flowers of the section Phlox have long styles bringing the stigma to the orifice of the tube, and the anthers also stand close to the orifice. In the sections Divaricatae and Occidentales, by contrast, the stigma and anthers are included deep within the corolla tube. The floral mechanism in the section Phlox transfers pollen on the tongue base or face of the Lepidopteran visitor, whereas the floral mechanism in the other two sections brings about tongue-tip pollination.

One advantage of tongue-tip pollination over tongue-base pollination in a Lepidoptera flower is that it makes better use of a variety of visitors with diverse tongue lengths. Even though the proboscis of some Lepidopteran visitor is much longer than the corolla tube, so that the head stands away from the flowers, the proper contacts are still made by the tongue tip which cannot avoid passing by the pollen and stigma en route to the nectar.

Now long-tongued hawkmoths are attracted to the flowers of the section Phlox and the section Divaricatae alike. Since the corolla tube is often much shorter than the tongue of the sphingid, the head of the moth stands back from the flowers. Such visits do not normally lead to pollination of floral mechanisms of the type found in section Phlox, but are fully effective in the floral mechanisms of the Divaricatae which utilize the tip of the moth's proboscis.

SECTION PHLOX

The 20 species of perennial herbs belonging to this section have numerous showy flowers, which in most species are clustered in round-topped terminal inflorescences. The corolla attains its maximum lengths for the genus in the section Phlox. The stamens and style are long, as noted earlier, and the anthers and stigma stand at or slightly above the orifice to the tube.

Phlox glaberrima   *Fig.* 7

This perennial herb occurs in moist open woods and grasslands of the eastern United States. The plants bloom in spring and early summer. The purple, pink, or white flowers are grouped in round-topped panicles and have a faint fragrance. The salverform corolla contains nectar at the base of a tube 2 cm. or less long and bears anthers and stigma around the orifice. The plants are self-incompatible (Levin, 1963).

Robertson (1895; 1928, 151) observed that the flowers in a population near Carlinville, Illinois, were visited and pollinated by Monarch, Swallowtail, and Sulphur butterflies, and by a skipper (Polites) and a small diurnal moth (Scepsis). The proboscis lengths of the butterflies correspond well with the corolla tube length of the Phlox, so that the butterflies can extract nectar successfully from these flowers. The basal part of the proboscis probably picks up pollen in the course of feeding.

FLORAL MECHANISM. The purple, pink, or white flowers are grouped in terminal round-topped panicles and have a faint fragrance. In a population of *P. glaberrima interior* observed by Robertson (1895) the flowers were purple, the corolla tube 1.6 to 1.8 cm. long, and the limb 2 cm. in diameter. Anthers and stigma lie at or near the orifice and, despite a slight protandry, overlap in period of maturity so that self-pollination is possible, though ineffective as shown below.

BREEDING SYSTEM. Levin (1963) has recently reported that *Phlox glaberrima interior* is self-incompatible.

INSECT VISITORS. Robertson (1895, 1928) made observations in a natural population growing on the prairie near Carlinville, Illinois. This population is now treated as *P. glaberrima*

Fig. 7. Phlox glaberrima and Colias philodice (Pieridae). Life size.

interior. Robertson recorded several kinds of butterflies, skippers, and moths pollinating the flowers, and a syrphid fly feeding on pollen but not pollinating. The modern equivalents of Robertson's names are listed below.

LEPIDOPTERA

Papilionidae: *Papilio cresphontes. Papilio philenor. Papilio ajax. Papilio thoas.*

Other families: *Danaus plexippus* (Danaidae). *Colias philodice* (Pieridae). *Polites peckius* (Hesperiidae). *Scepsis fulvicollis* (Syntomidae).

DIPTERA

Syrphidae: *Syrphus americanus.*

Phlox paniculata

This perennial herb occurs naturally in the eastern United States and is cultivated widely as a garden plant. It blooms in summer and fall, producing dense hemispherical inflorescences composed of fragrant pink, purple, or white flowers. The salverform corolla has a tube about 2 cm. long containing nectar at the base and bearing the anthers and stigma at or near the orifice. The flowers are protandrous.

In gardens the plants attract butterflies, such as the swallowtail, *Papilio rutulus,* which perches on the inflorescence and inserts its 2 cm. long proboscis into the corolla tube for nectar, and Hemaris and other sphingids with tongues of similar length which feed from a hovering position. These insects pick up and carry pollen on the base of the proboscis or face. Limited observations in nature, which should be extended, indicate that the species is pollinated by butterflies in natural habitats (Wherry, 1933). Various flies and bees have been recorded as casual visitors to the flowers of garden plants in Germany.

FLORAL MECHANISM. The fragrant pink, purple, or white flowers are clustered in a massive panicle. The corolla tube varies from 1.8 to 2.4 cm. long, or exceptionally from 1.6 to 2.6 cm., according to Wherry (1955, 120), and averaged about 2 cm. long in two horticultural varieties grown by the senior author. The tube is 2 mm. wide at the orifice and narrows down below. There is a slight upward curvature in the tube. Nectar collects at the base and may rise a few millimeters up in the tube. A hairy zone near the base of the tube helps to spread the nectar out in a thin film. Some of the anthers stand at the orifice, others 4 or 5 mm. below. The long style when fully extended brings the stigma to the orifice. The flowers are protandrous, as Sprengel (1793) first noted.

INSECT VISITORS. Nearly all of the observations have been made on garden plants. Wherry (1933) has noted that in nature the flowers are pollinated by "numerous butterflies," without giving details. The available observations of cultivated plants are in agreement.

GERMANY (Sprengel, 1793, 105; Müller, 1883, 407; Knuth, 1909, 113; Hegi, 1927, 2113; Porsch, 1939, under the synonym *Phlox decussata;* Werth, 1956, 170)

Lepidoptera: Butterflies (Sprengel). *Macroglossa stellatarum* (Sphingidae). *Plusia gamma* (Noctuidae).

Bees: *Anthidium strigatum* (Megachilidae). *Anthidium flavipes* (Megachilidae). *Halictus smeathmanellus* (Halictidae).

Diptera: *Conops flavipes* (sucking nectar, Conopidae). *Eristalis tenax* (eating pollen, Syrphidae). *Echinomyia fera* (= Cnephaliodes, Tachinidae).

CLAREMONT, CALIFORNIA

Papilio rutulus (Papilionidae). *Hemaris senta* (Sphingidae).

NEW HAMPSHIRE (David P. Gregory, personal communication):
Hemaris gracilis (Sphingidae). *Dolba hylaeus* (Sphingidae).

Phlox maculata

Phlox maculata is a spring-blooming perennial herb of moist meadows and boggy ground in the northeastern United States. The bright purple, fragrant, salverform flowers are grouped into cylindrical panicles. The plants are selfincompatible (Levin, 1963) and are cross-pollinated by "various kinds of butterflies" (Wherry, 1932).

Phlox stolonifera

This mat-forming plant of the Appalachian Mountains produces its fragrant violet or purple flowers in the spring. Wherry (1931) has observed the flowers being visited by butterflies.

Phlox subulata

This low mat-like plant of the northeastern United States blooms in the spring. The purple, pink, or white flowers give off a sweet or a pungent fragrance (Wherry, 1955, 73). The flowers are protandrous and are apparently selfcompatible (Knuth, 1909, 113).

It is of historical interest to note that Gray (1870, 248) reported heterostyly in *Phlox subulata* as a result of lumping it with another species, *Phlox nivalis*, which has short styles. Darwin received a mixture of material of the two species from Gray, on the basis of which he discussed the possibility of heterostyly, concluding that "the whole case is perplexing in the highest degree" (Darwin, 1877).

Wherry (1955, 73) states that the flowers are visited by butterflies and beeflies.

Phlox stansburyi and P. dolichantha   *Plate III D, Fig. 8*

Phlox stansburyi is a perennial herb of sagebrush or juniper-covered slopes in the Great Basin and *P. dolichantha* is a related endemic form in the San Bernardino Mts. of southern California. The erect white flowers have a long tube which is 3 cm. long in many races of *P. stansburyi* and 3 to 4 cm. long in *P. dolichantha*.

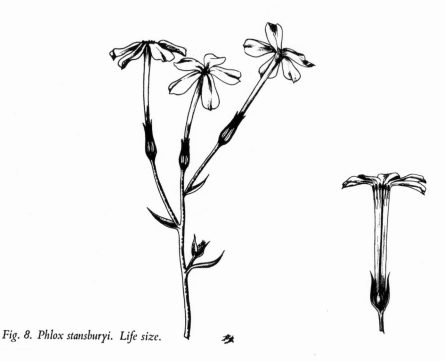

Fig. 8. Phlox stansburyi. Life size.

A strain of *P. stansburyi* is probably but not certainly self-incompatible on the basis of preliminary tests. The characteristics of the flowers strongly suggest hawkmoth pollination.

SECTION DIVARICATAE

This section includes some 20 species of woody-based plants, perennial herbs, and annuals in the United States and Mexico. The style is only a few millimeters long and the stigma is thus borne deep within the corolla tube. The anthers are also included within the corolla tube, usually at different levels above the stigma.

Phlox divaricata *Plate III A*

This spring-blooming spreading perennial occurs in woods in the Mississippi basin and the eastern states. The violet, honeysuckle-scented flowers are grouped in panicles. The corolla tube is about 1.5 cm. long and contains the stigma at its base and the anthers on its inner wall. Both the nectar and the pollen are out of easy reach of many flower-visiting insects. Long-tongued insects can reach the nectar and in so doing may carry pollen on the distal part of the proboscis to a stigma.

The common flower visitors of a population at Carlinville, Illinois, were butterflies, skippers, diurnal moths, bumblebees and some other bees, and a beefly (Robertson, 1891, 1895, 1928). The bees and beeflies may reach the upper nectar but may or may not contact the deeply buried stigma. The chief agents of proboscis pollination, by virtue of their longer tongues, are the Lepidoptera.

FLORAL MECHANISM. The violet flowers are grouped in panicles and give off a honeysuckle-like fragrance (Wherry, 1955, 39 ff.). The corolla tube is 1.1 to 1.8 cm. long in the species as a whole and 1.4 or 1.5 cm. long in the Illinois population for which pollination records are available. The stigma is deeply buried inside the tube and the anthers are inserted on the inner corolla wall.

INSECT VISITORS. Robertson (1891, 1895, 1928) observed *Phlox divaricata laphami* and its visitors near Carlinville, Illinois. Many of the recorded species of visitors were seen repeatedly in different years. Modern equivalents are used below for Robertson's Lepidoptera.

LEPIDOPTERA

Papilionidae: *Papilio troilus. Papilio cresphontes. Papilio marcellus. Papilio glaucus. Papilio philenor. Papilio ajax. Papilio thoas.*

Danaidae: *Danaus plexippus.*

Pieridae: *Colias philodice.*

Hesperiidae: *Epargyreus tityrus. Polites themistocles. Erynnis icelus. Thorybes bathyllus. Poanes zabulon. Poanes hobomok.*

Sphingidae: *Hemaris thysbe. Celerio lineata.*

Phalaenidae: *Autographa falcifera simplex.*

HYMENOPTERA

Apidae: *Bombus americanorum. Bombus consimilis. Bombus impatiens. Bombus ridingsii. Bombus separatus. Bombus vagans. Bombus virginicus. Psithyrus variabilis. Tetralonia dilecta.*

Megachilidae: *Osmia cordata.*

DIPTERA

Bombyliidae: *Bombylius atriceps.*

Phlox pilosa

This species of perennial herb, like the related *P. divaricata*, has a wide distribution in the eastern United States, where it inhabits open woods and meadows and blooms in spring and early summer. The fragrant purple or pink flowers are grouped in open panicles. The salverform corolla has a limb 2 cm. broad and a tube 1.0 to 1.5 cm. long in two populations in Illinois and Missouri for which pollination records are available. The nectar and stigma lie at the base of this tube and the anthers in the middle portion.

The insects which are best fitted to obtain the well-buried nectar and to pollinate the deeply buried stigma are Lepidoptera with long tongues. Various butterflies, skippers, and moths have been observed as frequent flower visitors in both Illinois (Robertson, 1895, 1928) and Missouri. In the Missouri population the butterflies were perching on the flowers to feed on the nectar and were carrying pollen on the distal part of their long tongues. A bumblebee passed by the Phlox plants and investigated the flowers but rejected them and flew on to nearby Delphinium flowers which it did visit. In the Illinois population, on the other hand, Robertson did record legitimate visits by bumblebees and beeflies. But the proboscis of these insects is probably not as well suited by its length and structure to carry pollen to the stigma as is the broad moist tip of the long Lepidopteran tongue.

FLORAL MECHANISM. In the races studied from the standpoint of pollination, the flowers are grouped in open panicles, are fragrant, sometimes with a scent of cloves (Wherry, 1955), and are purple or pink, often with a contrastingly colored eye spot in the center of the limb. The corolla limb is 2 cm. broad, and the corolla tube 1.0 to 1.5 cm. long in the Carlinville population studied by Robertson and 1.5 cm. long in the Ozark populations observed by ourselves. The stigma is at the base of the corolla tube and the anthers are disposed on three levels in the midportion of the tube.

INSECT VISITORS. The Lepidoptera observed by Robertson are listed below under their modern names. Wherry (1955, 49) has also observed small butterflies visiting the flowers of the prairie race of this species.

CARLINVILLE, ILLINOIS (Robertson, 1895, 1928)
  Papilionidae: *Papilio troilus. Papilio glaucus. Papilio ajax.*
  Nymphalidae: *Vanessa virginiensis. Phyciodes tharos.*
  Pieridae: *Colias philodice.*
  Lycaenidae: *Lycaena thoe.*
  Hesperiidae: *Epargyreus tityrus. Polites peckius. Thorybes bathyllus. Thorybes pylades.*
  Phalaenidae: *Autographa falcifera simplex.*

Apidae: *Bombus americanorum. Bombus auricomus. Bombus griseocollis. Bombus impatiens. Anthophora ursina. Tetralonia dilecta. Nomada superba.*

Bombyliidae: *Bombylius atriceps.*

ST. JAMES, MISSOURI (V. G. and K. G.)
Papilionidae: *Papilio troilus.*
Nymphalidae: *Vanessa sp. Speyeria cybele.*
Pieridae: *Colias eurytheme.*

Phlox nana

Phlox nana is a low tufted perennial of open woods in New Mexico and northern Mexico. The purple, pink, or white flowers are faintly fragrant and have a tube 1.2 to 1.8 cm. long. Butterfly pollination has not yet been observed in this species, but is to be expected on the basis of the floral characters. Cockerell (1902) saw a short-tongued bee, *Agapostemon texanus* (Halictidae), trying repeatedly but unsuccessfully to extract nectar from the flowers in a population near Las Vegas, New Mexico.

Phlox drummondii

Phlox drummondii is an annual herb of moist grassland and open oak woods on the Texas coastal plain. The flowers are bright red or other hues and slightly fragrant and are oriented upward and outward in few-flowered cymes. The corolla tube is about 1.5 cm. long, broad, and slightly curved. The stigma and nectar lie at the base of this tube and the anthers on the upper wall. The plants are essentially self-incompatible and predominantly outcrossing in the case of most races of the species (Fryxell, 1957; Erbe and Turner, 1962).

We have observed the hawkmoth, *Celerio lineata*, feeding on the flowers from a hovering position, and effecting tongue-tip pollination, in a population in Gonzalez County, Texas, while Erbe and Turner (1962) have reported occasional visits by butterflies.

FLORAL MECHANISM. The bright-colored, fragrant, salverform flowers are borne in few-flowered cymes and point upward and outward. A colony near Palmetto State Park, Gonzalez County, Texas, consists predominantly of plants with bright red flowers, intermixed with which are rare individuals with purplish or white flowers. The corolla tube in these plants is broad, slightly curved, and about 1.5 cm. long, and the limb is 2.0 to 2.5 cm. broad. The distance from the orifice of the tube to the deep-seated stigma is 1.2 or 1.3 cm. The anthers are disposed on the upper corolla tube wall.

BREEDING SYSTEM. Fryxell (1957, 198) reports that plants grown by E. Paterniani in Brazil were partially self-incompatible and had at least 80% natural outcrossing. Erbe and Turner (1962) have found that several races are self-incompatible, but one exceptional race, *P. drummondii littoralis*, is partially self-compatible and autogamous.

INSECT VISITORS. In a natural population at the edge of post oak woods near Palmetto State Park, Gonzalez County, Texas, during the daytime in April 1955, the senior author saw the common hawkmoth, *Celerio lineata*, feeding on the flowers from a hovering position and systematically covering the entire population from one end to the other, a distance of 200 feet. The hawkmoth was carrying pollen on its tongue tip, which must contact the stigma in probing for nectar, and the flowers visited by the moth were found to have pollen on the stigmas.

Erbe and Turner (1962) report that the flowers are visited occasionally by butterflies, probably *Papilio philenor*, and cite similar unpublished observations by Whitehouse.

Phlox cuspidata and P. roemeriana

These two annual species of Texas are related to *P. drummondii*. Erbe and Turner (1962) found strains of *P. cuspidata* to be self-incompatible, though the race *P. cuspidata humilis* is highly autogamous. Wherry (1955, 36) observed butterflies visiting the flowers of *P. roemeriana*.

SECTION OCCIDENTALES

This section contains over 20 species of woody-based, needle-leaved, cespitose or cushion-like subshrubs. The group is widespread in the mountains and high plains of western North America; two species occur in Alaska, and one in Asia. The flowers are terminal on the branches and solitary or few in number. The development of numerous parallel flowering shoots, each with one or a few flowers, leads, however, to as close an aggregation of flowers as is found in the herbaceous Phloxes with compact panicles. The corollas are somewhat shorter on the average than in other sections of the genus. The stigma is always included within the corolla tube, though not as deeply as in the section Divaricatae.

Phlox caespitosa   Fig. 9

This mat-like plant occurs throughout the Rocky Mt. region. In a population which we observed in alpine tundra in the Front Range of Colorado, the plants begin to bloom almost as soon as the snow has melted off the mats. The white

Fig. 9. Phlox caespitosa and Euxoa messoria (Noctuidae). Life size.

flowers form a nearly solid mass and are open and fragrant at dusk and in the night. The corolla tube is 1 cm. long.

Large numbers of noctuid moths, *Euxoa messoria*, with a proboscis 1 cm. long, were observed feeding on the flowers by night. The moths carry pollen on the distal half of the proboscis from flower to flower.

FLORAL MECHANISM. In a population studied in Rocky Mt. National Park, Colorado, the plants begin to bloom almost as soon as the snow has melted off the mat-like plant body, producing a solid or nearly solid mass of flowers. The flowers are white with a pale violet tinge and are open and fragrant at dusk and in the night. The corolla tube is 1 cm. long, the stigma is about 5 mm. below the orifice, and the anthers lie between the stigma and orifice.

INSECT VISITORS. No insect visitors were seen during the daytime on five days of observation in 1961. But on a mild evening in June 1962 large numbers of moths, *Euxoa messoria* (Noctuidae), were feeding on the flowers. The moths fly to a patch of flowers, land, and feed by inserting their 1 cm. long proboscis into the corolla tube for nectar. Having sucked the nectar from one flower, they then *walk* over the low mats to other flowers in the same patch, and eventually fly away to a new patch. In inserting the pollen-covered proboscis into a corolla tube they must contact the stigma and effect cross-pollination.

Phlox diffusa

This species of the Rocky Mts. and Pacific slope is generally similar to *Phlox caespitosa* in having a cespitose habit and a mass of fragrant pink, lavender, or white flowers. In a population observed on Mt. Dana in the Sierra Nevada, the corolla tube is 1.2 cm. long, and the stigma stands 8 mm. below the orifice. A noctuid moth, *Heliothis obsoleta*, was feeding on these flowers by day in August 1956 and was bringing about tongue-tip pollination. Several bumblebees working meanwhile on flowers of *Lupinus lyallii* intermixed with the Phlox paid no attention to the latter. In another population of *Phlox diffusa* in Ebbetts Pass farther north we saw skippers (unidentified) visiting and pollinating the flowers.

Phlox multiflora and P. andicola

The first species occurs in the northern Rocky Mt. area and the second in the northern Great Plains. The flowers are similar to those of the preceding species, with corolla tubes about 1 cm. long, included stigmas, and a marked fragrance. Wherry (1955, 150) has observed small butterflies visiting the flowers of *Phlox andicola*. The senior author saw a Monarch butterfly, *Danaus plexippus*, feeding on and pollinating *Phlox multiflora* in the Uintah Mts. of Utah in 1955.

MICROSTERIS

Microsteris is a satellite genus close to Phlox consisting of the small annual autogamous species, *Microsteris gracilis*, which occurs widely in western North America and temperate South America. The small inconspicuous flowers set seeds autogamously, and the inbred progeny of several strains tested in Claremont are fully vigorous. Nevertheless, insects do visit the flowers and cross-pollinate the plants occasionally. Unidentified small bees have been observed visiting the flowers on Mt. Pinos, Ventura County, California.

5
GILIA

The genus Gilia contains about 67 species of spring-blooming herbaceous plants, mostly annuals but including some perennials and biennials, inhabiting the plains, mountains, and deserts of Texas, Mexico, the west, and southern South America. The group is heterogeneous, exhibiting diverse trends in vegetative and floral characters both between and within the five sections, so that it is difficult to generalize about the genus as a whole. This heterogeneity is also true of the mode of pollination. The flowers are usually funnelform, but sometimes campanulate or nearly salverform, and commonly much reduced in size, and are pollinated by a wide variety of agents.

SECTION GILIASTRUM

The section Giliastrum consists of perennial or annual herbs with blue campanulate or rotate flowers. Unfortunately, this group is poorly known from the standpoint of pollination ecology.

Gilia rigidula and *G. foetida*   Plate I D, Fig. 10

Gilia rigidula and *G. foetida* are sprawling perennial herbs which occur on the steppes of North and South America respectively. The large showy fragrant flowers are campanulate with a broad purple or blue limb and a short nectar-containing tube. The orange anthers and the stigma are well exserted. Strains of *G. rigidula* from Carta Valley, Texas, and of *G. foetida* from Uspallata, Argentina, have been tested in Claremont and found to be self-incompatible. The floral characters are such as to suggest bee pollination, but field observations are required to verify this supposition.

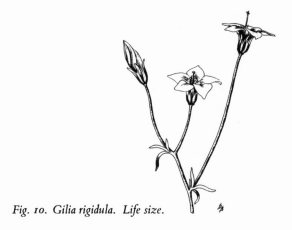

Fig. 10. Gilia rigidula. Life size.

Gilia incisa and G. stewartii

These two species of annuals of Texas and Mexico are fairly closely related to *G. rigidula*. Their flowers are similar in general features but smaller in size and paler in color than those of *G. rigidula*. *Gilia incisa* from Austin, Texas, is self-compatible and at least partially autogamous. The results of tests with *G. stewartii* from San Luis Potosí, Mexico, were conflicting but suggest that this plant is also self-compatible.

SECTION GILIANDRA

This section comprises several species of perennial and biennial herbs in the Rocky Mt. and intermountain regions and several annual species in the western deserts. The flowers are funnelform with usually well exserted stamens; they tend to be large and showy in the perennials and biennials and reduced and autogamous in the desert annuals.

Gilia pinnatifida

This small biennial herb of the Rocky Mts. has pale blue flowers with a yellow eye. These are oriented to the side or canted slightly upward and have long exserted stamens and style on the lower side. The nectar is contained in a narrow corolla tube about 4 mm. long. Bumblebees, mainly *Bombus flavifrons* and *B. centralis*, but also *Bombus melanopygus* and the megachilid bee *Megachile melanophaea*, have been observed feeding on the flowers and effecting pollination in three populations in Colorado. The bees cling onto the exserted stamens to collect pollen or probe for nectar and carry pollen on their venter.

FLORAL MECHANISM. The pale blue flowers form a loose open inflorescence and lie horizontally or canted slightly upward. In one population studied, the following measurements were obtained. The blue-violet corolla limb is 8 or 9 mm. broad; the orifice to the tube is 1 mm. broad and is outlined with yellow spots; and the tube is 4 mm. long. The blue stamens are exserted on the lower side of the corolla, the anthers standing 4 or 5 mm. beyond the orifice, and the stigma is exserted beyond the anthers.

INSECT VISITORS. Populations have been observed at three sites in Colorado: Ward, Boulder County; Boulder Canyon, Boulder County; and Morrison, Park County. Bumblebees were the most common visitors and efficient pollinators in each population. *Bombus flavifrons*, *Bombus melanopygus*, and *Megachile melanophaea* were seen in the Ward population; and *Bombus centralis* in the Boulder Canyon and Morrison populations.

Several individuals of *Bombus flavifrons* were working simultaneously in the Ward population. One individual was collecting pollen and the others nectar. The latter cling to the exserted stamens and insert the extended proboscis, which is about 6 mm. long, into the corolla tube, and as they move through the population they carry pollen on the venter. The behavior of *Bombus centralis* in the other populations was similar.

Gilia pentstemonoides

This narrow endemic species from Montrose County, Colorado, has flowers similar to those of *G. pinnatifida.* We have propagated plants in Claremont from transplants collected by Dr. Wm. A. Weber near Blue Mesa. These plants are self-incompatible.

The Gilia leptomeria group

Gilia leptomeria, hutchinsifolia, and *micromeria* are annual species of the western deserts with reduced self-pollinating flowers and autogamous reproduction.

LEAFY-STEMMED GILIAS

The section Gilia, or Leafy-stemmed Gilias, is a group of nine species of annual herbs with blue funnelform flowers which occurs along the Pacific coast of North America and again in temperate South America.

Gilia tricolor   *Fig.* 11

Gilia tricolor is a common and showy annual of the foothills and valleys of central and northern California. The broad funnelform flowers are blue-violet

Fig. *11*. *Gilia tricolor, Chelostomopsis rubifloris* (*Megachilidae, above*),
and *Osmia distincta* (*Megachilidae, below*). *Life size.*
(*The bees are shown enlarged in Fig. 30 C, F.*)

and orange with purple spots in the throat. They lie canted to the side with the
stamens exserted to varying distances on the lower side. The plants are self-
compatible. The flowers are visited most abundantly by various small megachilid,
halictid, and andrenid bees, which settle on the stamens or in the throat to collect
pollen or obtain nectar. Various beeflies also visit the flowers frequently. Both
the bees and beeflies carry pollen on the venter.

FLORAL MECHANISM. The broadly funnelform flowers are erect and canted to the side.
The corolla limb is blue-violet, the short tube and lower throat are yellow to orange, and the
expanded throat is decorated with five pairs of purple spots. The throat and limb are larger in
the northern than in the southern race of the species, the throat being typically 5 mm. wide in the
former and 3 mm. wide in the latter. The stamens are exserted on filaments of varying length.
As the flower lies tipped to the side, the two lowermost stamens are relatively long and upcurved,

the two lateral stamens are of medium length, and the upper one is short. The anthers dehisce several days before the stigma is mature. The mature stigma then extends 1 or 2 mm. beyond the anthers.

BREEDING SYSTEM. Brand (1905) found that *Gilia tricolor* grown in Berlin is self-compatible, and Eastwood (1893) expressed the same conclusion on field studies. We have caged and artificially self-pollinated three strains and found them to be self-compatible (Grant, 1952b). Some capsules form autogamously on the caged plants, especially toward the end of the flowering season.

INSECT VISITORS. The following stations are all in California.

KAWEAH RIVER, TULARE COUNTY (V. G. and K. G.)

Bees: Halictus subgenus Halictus (Halictidae). *Andrena sp.* (Andrenidae). *Osmia distincta* (Megachilidae). *Chelostomopsis rubifloris* (Megachilidae). *Tetralonia sp.* (Apidae).

Flies: *Bombylius major* (Bombyliidae). Conopidae.

Butterfly: *Colias eurytheme* (Pieridae).

WILLIAMS, COLUSA COUNTY

Dufourea sp. (Halictidae). *Chloralictus sp.* (Halictidae). *Bombylius lancifer* (Bombyliidae). *Malachius mirandus* (Melyridae).

WOODEN VALLEY, NAPA COUNTY

Dufourea sp. (Halictidae).

MIDDLETOWN, LAKE COUNTY

Dufourea sp. (Halictidae).

DAVIS, YOLO COUNTY (Jack Hall, personal communication)

Bombyliidae: *Bombylius lancifer. Bombylius major. Pantarbes sp. Conophorus sp.*

Bees: unidentified andrenids and halictids.

MADERA COUNTY (Hurd and Michener, 1955)

Chelostoma incisulum (Megachilidae).

Gilia angelensis

This spring-blooming annual of the valleys of southern California is related to *Gilia tricolor* and differs from that species in having smaller and less colorful flowers. The plants are self-compatible and autogamous in the absence of insects. The flowers are visited and pollinated by various kinds of insects, however, and the protandry and other features of the floral mechanism probably promote outcrossing by such visitors.

The following classes of flower-visiting insects have been observed: small halictid and megachilid bees, honeybees, syrphid flies, melyrid beetles, and butterflies and skippers.

The small solitary bees land in the flowers and enter the small corolla throat, which they fit into well, to obtain pollen or nectar, and in either case become dusted with pollen on various parts of the body. These bees are by far the most effective pollinating agents.

The honeybees are too large and heavy for these Gilia flowers, but nevertheless they bring about considerable pollination. The syrphid flies feed on the nectar of the Gilia flowers actively at certain seasons, and as the pollen sticks to their body hairs to some extent they are somewhat effective as pollinators. The melyrid beetles bring about a limited amount of pollination. The butterflies and skippers have not been observed to carry any pollen.

We kept one population in Pomona Valley under close and frequent observation during the spring of 1962. The blooming season in this population passed through four main periods. (1) The initial period (March 25 to April 4), when the plants were first beginning to flower, was marked by cool weather. (2) In the early period of full bloom (April 4 to 10), the weather was mild, with temperatures sometimes as high as 96° F. but usually less. (3) The period of full flowering in midseason (April 13 to 18) occurred during generally warmer weather with some hot days when the temperature rose to 108° F. (4) The end of the flowering season (after April 18) found only a few straggling flowers in bloom.

Activity on the part of native insects was rare or absent during the initial and late periods. Honeybees were moderately active in the initial period, but later, when other species of flowers appeared, transferred their attentions to them and were seen more seldom on the Gilia. In the early period of full bloom under mild temperatures, syrphid flies were the most common visitors and chief pollinating agents. In the middle period of full bloom with warm to hot temperatures, the syrphid flies were less in evidence, except in the cool hours of early morning, whereas the native solitary bees took over the role of predominant visitors and pollinators. The spectrum of pollinating agents is thus subject to seasonal changes and, in midseason, to hourly changes through the day.

FLORAL MECHANISM. The flowers are like those of *Gilia tricolor* but smaller and less colorful. The corolla is light blue-violet to white and without spots in the throat. The corolla throat is 3 mm. wide and the tube 2 mm. long. The anthers stand at the orifice of the throat and the stigma extends beyond them. The flowers, which are closed at night by the folding of the corolla lobes, open during the warm sunny hours of the morning and close again in late afternoon before sundown.

BREEDING SYSTEM. A strain from Perris, Riverside County, California, has been tested and found to be autogamous when isolated from insects. The I_1 progeny were fully vigorous (Grant, 1952b).

INSECT VISITORS. We observed a population in Pomona Valley, Los Angeles County, California, at frequent intervals all through the blooming season of 1962. Supplementary observations are available for two other populations.

POMONA VALLEY, LOS ANGELES COUNTY

Bees: Halictus subgenus Seladonia (Halictidae). Halictus subgenus Halictus (Halictidae). *Chloralictus sp.* (Halictidae). *Chelostoma sp.* (Megachilidae). *Apis mellifera* (Apidae). Syrphidae: *Eristalis tenax. Syrphus americanus.* Two unidentified medium-sized syrphids. Lepidoptera: *Vanessa carye* (Nymphalidae). *Hylephila phylaeus* (Hesperiidae). Beetles: Melyridae.

TEMESCAL CANYON, RIVERSIDE COUNTY

Dufourea sp. (Halictidae). *Apis mellifera.*

SAN JACINTO MTS., RIVERSIDE COUNTY (Hurd and Michener, 1955, 188)

Ashmeadiella californica (Megachilidae).

In all populations the small solitary bees are the most effective pollinating agents. Syrphid flies are somewhat effective as pollinators, and so to a much lesser degree are the occasional melyrid beetles. The butterflies and skippers observed in this study did not contact the anthers well and were not carrying pollen.

Gilia achilleaefolia Fig. 12

This species of annual herb occurs in the live-oak savanna of the South Coast Ranges in California. The plants have blue-violet funnelform flowers and are self-compatible. Otherwise the species is highly variable and consists of numerous local races differing in floral mechanism, breeding system, and mode of pollination.

Races occurring on sunny hillsides in San Luis Obispo and Santa Barbara counties have capitate heads composed of large, bright blue-violet flowers with full throats 6 or 7 mm. wide. The flowers are non-autogamous in early spring but become predominantly autogamous at the end of the blooming season. The I_1, I_2, and I_3 progeny exhibit considerable inbreeding depression which decreases in successive generations. Progeny tests of population plants indicate that out-crossing prevails, though some selfing also occurs, under natural conditions of pollination at the height of the blooming season (Grant, 1954a).

The chief pollinators of these races are halictid and andrenid bees of medium size, that is, of a size somewhat smaller than honeybees. These bees enter the expanded corolla throat, which they fit into very well, for purposes of extracting nectar or pollen and in so doing pick up pollen on various body parts. Beeflies (*Bombylius lancifer*) are also effective pollinators.

Races occurring in sunny habitats farther north have bright-colored but smaller flowers with a narrower corolla throat, which is only 3 mm. wide in a typical example. These flowers are pollinated by small solitary bees, such as Dufourea, which fit into the small throat, and also by Bombylius. There is

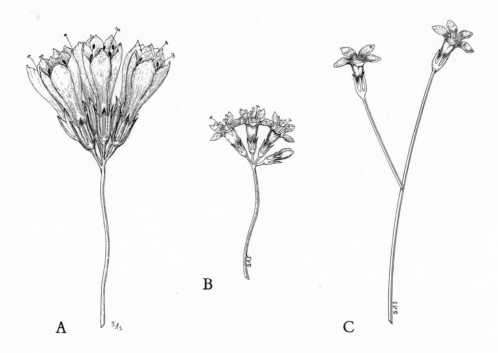

A   B   C

Fig. 12. Gilia achilleaefolia. Life size.

(A) San Luis Obispo race.   (B) Pinnacles race.   (C) Shaded woods race.

experimental evidence that the northern sun-loving races are capable of more regular autogamy in the absence of insect visits than are the southern races.

Finally, some races inhabiting shady woods have small, pale-colored, non-capitate flowers. These plants are predominantly autogamous and produce fully vigorous inbred progeny. Their flowers are occasionally visited and cross-pollinated by small bees, however.

FLORAL MECHANISM. This species is very variable in floral characters, ranging from races with large flowers in heads to races with small solitary flowers. The large-flowered capitate races are treated collectively as *Gilia achilleaefolia achilleaefolia* and the small-flowered non-capitate races as *G. a. multicaulis.*

Let us consider first the floral mechanism in *G. a. achilleaefolia*. Here the heads are fan-shaped and consist of between 8 and 25 blue-violet funnelform flowers. In the populations of San Luis Obispo and Santa Barbara counties the corolla throat is very ample and long, whereas the populations farther north have a smaller throat. For example, the diameter of the throat is 6 or 7 mm. in a population at Pt. Sal, Santa Barbara County, but 3 mm. in a population in Pinnacles National Monument, San Benito County. The anthers stand at different levels around the entrance to the throat. The stigma ripens after the anthers and is exserted 1 or 2 mm. beyond them. This spatial separation of anthers and stigma is more complete in the flowers of early spring than in those of late spring.

The flowers of *Gilia achilleaefolia multicaulis* are smaller, more weakly colored, and are often more or less solitary on long pedicels. In these forms the stigma and anthers occupy the same level from the beginning of flowering and regularly undergo self-pollination.

BREEDING SYSTEM. *Gilia achilleaefolia* is self-compatible as shown by tests of eight strains representing the main racial variations (Grant, 1954a).

Gilia achilleaefolia multicaulis is autogamous. Some of the smaller flowered races of *G. a. achilleaefolia* also set capsules freely in breeding cages.

The situation in the large-flowered San Luis Obispo populations of *G. a. achilleaefolia* is more complex. The anthers and stigma are well separated in the first flowers of early spring but not in the flowers produced at the end of the blooming season. It would be predicted from the observed seasonal changes in the floral mechanism that the early flowers are dependent on insect visits for pollination, whereas the late flowers are partially autogamous. To test this point, sister individuals were caged at different seasons, one in spring and the other in summer. The plant isolated in the spring set only one capsule on 67 flowers, while the plant isolated during the summer produced numerous capsules and only a few empty calyces (Grant, 1954a).

The vigor of the inbred progeny also differs as between the small-flowered and large-flowered races. The I_1 progeny of the autogamous *G. a. multicaulis* are fully vigorous, as would be expected.

The first inbred generation produced by selfing normal individuals of the San Luis Obispo strain of *G. a. achilleaefolia*, on the other hand, consisted largely of weak or inviable plants. Some of these died in the seedling stage, others developed into aberrant and male-sterile individuals, and still others were weaklings which never attained full development. Only a small proportion of the I_1 plants had normal vigor. Inviable and semi-viable types continued to be segregated, though with diminishing frequency, in the I_2 and I_3 generations derived from the exceptional normal I_1 individuals (Grant, 1954a).

Progeny grown from seeds harvested on open pollinated plants in the natural population of the same San Luis Obispo strain also included a few runts, but consisted predominantly of vigorous individuals. Under natural conditions of pollination, therefore, outcrossing apparently predominates over selfing in this race (Grant, 1954a).

INSECT VISITORS. *Gilia achilleaefolia* has been studied at four stations in the California Coast Ranges. The Pt. Sal and San Luis Obispo populations represent the full-throated race of *Gilia achilleaefolia achilleaefolia;* the Pinnacles population is a smaller flowered race of *G. a. achilleaefolia;* and the Tassajara population represents *G. a. multicaulis*.

PT. SAL, SANTA BARBARA COUNTY

Augochlora pomoniella (Halictidae). *Halictus farinosus* (Halictidae). *Andrena sp.* (Andrenidae). *Apis mellifera* (Apidae).

SAN LUIS OBISPO, SAN LUIS OBISPO COUNTY

Andrena sp. (Andrenidae). *Bombylius lancifer* (Bombyliidae). *Vanessa carye* (Nymphalidae). *Dione vanillae comstocki* (Nymphalidae). *Papilio rutulus* (Papilionidae). Unidentified Melyridae. *Mordella sp.* (Mordellidae).

PINNACLES NATIONAL MONUMENT, SAN BENITO COUNTY

Dufourea sp. (Halictidae). *Bombylius sp.* (Bombyliidae). Unidentified Pieridae. *Adela simpliciella* (Adelidae). *Eschatocrepis constrictus* (Melyridae).

TASSAJARA SPRINGS, MONTEREY COUNTY

Hylaeus sp. (Colletidae).

Gilia capitata   Plate I G

Gilia capitata is a polytypic species of fairly large annual herbs on the Pacific slope from Washington to southern California. The fragrant, blue-violet, funnelform flowers are grouped in dense spheroidal terminal heads. The stamens and styles are exserted above the corolla throat. This throat is slender in the mountain races and full in the races of sandy lowlands.

The mountain races tend to be self-incompatible and the sandy lowland races self-compatible. Outcrossing is probably promoted in the latter by two features of the floral mechanism, namely protandry and the elongation of the style beyond the stamens. Only in one exceptional isolated population has autogamy been observed at the end of the blooming season.

The flowers are regularly and actively visited during the warm part of the day by various bees, beeflies, beetles, and butterflies. These insects perch on the flowering heads to eat or collect the pollen or nectar. In the process, they contact the anthers and stigmas.

The butterflies obtain nectar from *Gilia capitata* but do not appear to pollinate the flowers to any significant extent, because the Gilia pollen does not normally adhere to those parts of their bodies which come into contact with the stamens. The melyrid beetles eat the pollen and carry more pollen on their bodies, but as they fly from plant to plant at relatively infrequent intervals they probably account for only a small amount of pollination.

The most effective pollinators of *Gilia capitata* are bees of various kinds—Andrenas, halictids, Osmias, bumblebees, honeybees, etc.—which carry much pollen on their bodies and fly actively and systematically from plant to plant. Beeflies (*Bombylius lancifer*) are also effective as pollinators in certain times and places.

FLORAL MECHANISM. This tall herb bears numerous flowers in spheroidal capitate heads at the ends of the branches. The individual flowers are blue-violet, fragrant, and funnelform. Nectar is produced at the base of the corolla tube. The stamens and styles are exserted above the corolla throat. The essential organs mature in protandrous order. Owing partly to the protandry and partly to the elevation of the stigma above the anthers, self-pollination does not normally occur automatically.

Although *Gilia capitata* is ordinarily protandrous, individual plants of poor vigor sometimes produce their styles precociously, while the corollas remain underdeveloped and the anthers abortive. The floral head in such cases has the aspect of a pincushion with a dozen or more long receptive styles protruding out of the unexpanded calyces. If the stigmas are pollinated artificially, normal seeds develop. It is possible that abnormal protogyny serves an adaptive role in this annual species, by enabling a plant living under unfavorable conditions to shorten the usual reproductive sequence and to produce a few seeds before it perishes (Grant, 1950, 272).

The size, color, and odor of the flowers are subject to geographical variation. In the northern and mountain races (G. c. *capitata* and G. c. *mediomontana*), the flowers are light blue-violet, sweet-scented, and small with a slender corolla throat and narrow linear lobes.

The races inhabiting sandy hills and plains along the coast and in the San Joaquin Valley (G. c. *chamissonis* and G. c. *staminea*) present the opposite extreme in floral characters. These races possess larger flowers with an expanded throat and broad oval lobes. The flowers are deep blue-violet and have a creosote scent (in G. c. *chamissonis*).

The extremes just mentioned are connected by intermediate forms, such as G. c. *pedemontana* in the Sierra Nevada foothills and G. c. *abrotanifolia* in southern California.

BREEDING SYSTEM. Ten strains representing all the main racial facies of *Gilia capitata* have been tested for self-compatibility and autogamy (Grant, 1950, 1952a). With one exception, none of these strains proved capable of autogamy. *Gilia capitata* requires insect visits for pollination. But the self-compatibility differs from race to race.

The broad-throated and broad-lobed races of sandy lowlands are self-compatible. (The strains tested are: G. c. *chamissonis*, Pt. Reyes, Marin County; *staminea*, Antioch, Contra Costa County; *pedemontana*, Kaweah River, Tulare County; *abrotanifolia*, Mentone, San Bernardino County.)

The slender-flowered mountain races, on the other hand, are self-incompatible. (The strains tested are: G. c. *capitata*, Mayacama Mts., Napa County; *tomentosa*, Tomales Bay, Marin County; *abrotanifolia*, Sequoia National Park, Tulare County; *abrotanifolia*, Kaweah River, Tulare County; *abrotanifolia*, Kernville, Kern County.)

This self-incompatibility is complete or nearly so in some strains. It is less strong in others where the selfing of numerous flowers leads to the production of a few capsules and seeds. Thus the self-incompatibility is very strong in G. c. *capitata* from the Mayacama Mts., but of reduced strength in G. c. *abrotanifolia* from the Kaweah River.

The reduced self-incompatibility of the Kaweah River race of G. c. *abrotanifolia* is evidently due to the expression of genetic factors of intermediate strength. The F_1 hybrids of this race with the almost completely self-incompatible Mayacama race of *G. c. capitata* are highly self-incompatible. The F_1 hybrids of *abrotanifolia* Kaweah × *staminea* Antioch, on the other hand, are about as self-compatible as their alternate parent (Grant, 1952a).

Autogamy has been mentioned above as a rare condition in *Gilia capitata*. A disjunct population of *G. c. tomentosa* which occurs as a probable Pleistocene relict on Mt. Diablo, Contra Costa County, sets seeds autogamously at the end of its flowering season.

INSECT VISITORS.

MAYACAMA MTS., NAPA COUNTY   (*G. c. capitata*)

Bees: *Dufourea sp.* (Halictidae). *Apis mellifera* (very abundant).

Diptera: *Bombylius lancifer* (Bombyliidae).

Lepidoptera: *Euphydryas chalcedona* (Nymphalidae). *Melitaea gabbii* (Nymphalidae). *Pieris rapae* (Pieridae). *Coenonympha californica* (Satyridae).

DARRAH, MARIPOSA COUNTY   (*G. c. mediomontana*)

Bees: *Andrena sp.* (Andrenidae). *Dufourea sp.* (very abundant) (Halictidae). Halictus subgenus Halictus (Halictidae). Halictus subgenus Seladonia (Halictidae). *Nomada sp.* (Apidae). *Bombus vosnesenskii* (mainly on Lupinus nearby) (Apidae). *Apis mellifera* (Apidae).

Beetles: Unidentified melyrid. *Mordella albosuturalis* (Mordellidae). *Trichodes ornatus* (Cleridae).

Lepidoptera: *Euphydryas chalcedona* (Nymphalidae). *Phyciodes myllita* (Nymphalidae). *Plebius icarioides* (Lycaenidae). *Colias eurytheme* (Pieridae). *Hesperia juba* (Hesperiidae).

KAWEAH RIVER, TULARE COUNTY   (*G. c. pedemontana*)

Bees: *Andrena prunorum* (Andrenidae). *Nomadopsis sp.* (Andrenidae). *Lasioglossum sisymbrii* (Halictidae). *Dufourea sp.* (Halictidae). *Agapostemon sp.* (Halictidae). *Ceratina sp.* (Apidae). *Bombus edwardsii* (Apidae). *Bombus vosnesenskii* (Apidae). *Apis mellifera* (Apidae).

Diptera: *Syrphus americanus* (Syrphidae).

Beetles: *Trichochrous suturalis* (Melyridae). *Mordella sp.* (Mordellidae). *Leptura lineola* (Cerambycidae). *Hoplia dispar* (Scarabaeidae). *Trichodes ornatus* (Cleridae).

Lepidoptera: Unidentified skipper.

KAWEAH RIVER, TULARE COUNTY   (*G. c. abrotanifolia*)

Bees: *Andrena sp.* (Andrenidae). *Nomadopsis sp.* (the same as on *G. c. pedemontana*) (Andrenidae). *Lasioglossum sisymbrii* (Halictidae). *Lasioglossum kincaidii* (Halictidae). *Ceratina sp.* (Apidae). *Apis mellifera* (Apidae).

Diptera: *Syrphus americanus* (Syrphidae). *Peleteria sp.* (Tachinidae).

Beetles: *Trichochrous suturalis* (Melyridae).

SAN GABRIEL MTS., LOS ANGELES COUNTY   (*G. c. abrotanifolia*)

Bees: *Andrena sp.* (Andrenidae). Halictus subgenus Seladonia (Halictidae). *Osmia sp.* (Megachilidae).

Diptera: *Bombylius lancifer* (Bombyliidae).

Beetles: *Anthaxia aenogaster* (Buprestidae).

Lepidoptera: *Euphydryas chalcedona* (Nymphalidae).

The autogamous species

Five species of Leafy-stemmed Gilias have small flowers which are automatically self-pollinating and produce fully vigorous inbred progeny. One of these species, *Gilia clivorum*, is a common element in the foothills of the California Coast Ranges (Grant, 1954b). The remaining four have a maritime distribution. *Gilia millefoliata* and *G. nevinii* occur on the coastal strand and offshore islands, respectively, of California, while the related *G. laciniata* and *G. valdiviensis* occur in Peru and Chile.

COBWEBBY GILIAS

The section Arachnion is the largest and taxonomically most complex section of the genus, consisting of about 28 known species, 25 of which will be considered below. The plants are spring-blooming scapose annuals which have their main center of distribution in the Mojave Desert, but occur widely throughout other desert areas of western North America and southern South America, and in the mountains and valleys west of the desert in California.

Gilia ochroleuca

Gilia ochroleuca occurs in southern California, where one geographical race, *G. o. bizonata*, inhabits piñon-juniper woodland in the mountains west of the desert, and the other race, *G. o. ochroleuca*, inhabits creosote-bush plains in the Mojave Desert. The funnelform flowers are borne on long slender pedicels in diffuse inflorescences and are self-compatible. Otherwise, the two geographical races differ in floral characters related to the mode of pollination and the breeding system.

The desert race with its small whitish flowers is predominantly autogamous. The western mountain race, *G. o. bizonata*, on the other hand, with larger flowers and long exserted styles, is non-autogamous and outcrossing to a considerable extent. The medium-sized, pinkish-violet flowers of *G. o. bizonata* are cross-pollinated by small solitary bees, most effectively by two species of Hesperapis (Melittidae) in one population observed.

FLORAL MECHANISM. *Gilia ochroleuca* consists of two geographical races, treated as subspecies, which are alike in some floral and inflorescence characters but different in others. The small-flowered race, *G. o. ochroleuca*, occurs in the creosote-bush zone (*Larrea divaricata*) of the

Mojave Desert; the large-flowered race, G. o. bizonata, in piñon-juniper woodland in the mountains west of the desert.

In the species as a whole the funnelform flowers are borne in pairs on moderately long pedicels of nearly equal length ($\pm$ 1 cm. long), and the inflorescence is consequently diffuse. The stamens are short, the anthers standing around the orifice of the corolla throat.

The flowers of the mountain race, G. o. bizonata, are medium-sized with a corolla limb 1 cm. broad, a throat 3 to 4 mm. broad at the orifice and 3 to 4 mm. deep, and a tube about 3 mm. long. The corolla is pinkish-violet with a yellow ring in the bottom of the throat, and the plant colony makes a bright splash of pinkish-violet color on the sandy ground. The two flowers of each pair bloom almost simultaneously, one opening about one day before the other. Each flower is protandrous, and the style at maturity is exserted well beyond the short stamens.

The flowers of the desert race, G. o. ochroleuca, are small, whitish, and self-pollinating.

BREEDING SYSTEM. Strains of both G. o. bizonata and G. o. ochroleuca have been found to be self-compatible. The latter race is in addition normally autogamous, and the inbred progeny are fully vigorous, whereas the woodland race, G. o. bizonata, is outcrossing to a considerable extent.

INSECT VISITORS. A population of G. o. bizonata in Cajon Pass, San Bernardino County, California, was observed on a warm spring day in 1963. Four species of small solitary bees were working on the flowers: Halictus sp. (Halictidae), Osmia sp. (Megachilidae), Hesperapis sp. (Melittidae), and Hesperapis sp. (another species) (Melittidae).

The bees settled down on the flowers and burrowed into the throat, some probing for nectar and others collecting pollen, and in either case getting pollen on their venters and legs. Being small and light, they could fit into the small corolla throat and were supported well by the long slender flower pedicels.

Of the bees mentioned above the two species of Hesperapis were by far the most effective pollinators, in terms of both numbers of individuals and individual effectiveness. The Halictus and Osmia were uncommon on the Gilia flowers, and the Osmia especially was working mainly on Lotus flowers nearby and strayed only occasionally to the Gilia. Other larger and heavier bees on the Lotus, too large and heavy for the Gilia flowers, were avoiding the latter consistently.

Beeflies (Bombylius lancifer) occasionally visit G. o. bizonata for nectar, as has been witnessed on Figueroa Mt., Santa Barbara County.

Gilia exilis

This species of the interior mountains of southern California is related to G. ochroleuca. The bluish flowers are funnelform with a medium-sized throat. A population from Dripping Springs, Riverside County, is known to be self-compatible. Hurd and Michener (1955, 32, 42, 64–65) have recorded three species of megachilid bees visiting the flowers of this plant in Mill Creek Canyon in the San Bernardino Mts., viz., Chelostoma cockerelli, Chelostomopsis rubifloris, and Hoplitis producta.

Gilia cana *Plate III G, K*

This polytypic species occurs in a series of geographical races throughout the mountains of the Mojave Desert and up the desert slopes of the Sierra Nevada. The fragrant pink or purple flowers form masses of color on the desert floor in years of favorable rainfall.

Two of the races, *G. c. triceps* and *G. c. speciosa*, have been tested and found to be self-compatible but non-autogamous and largely outcrossing. The latter race exhibits inbreeding depression. Exceptional autogamous individuals occur in certain polymorphic populations of *G. c. triceps*.

Most races of *Gilia cana* have slender funnelform flowers with tubes of varying length from 0.5 to 1.5 cm. long. The most consistent insect visitors and pollinators in five populations of the funnelform races have been small beeflies, chiefly Oligodranes, but sometimes Phthiria or Bombylius, which settle into the slender corolla throat to probe for nectar and pick up pollen on their bodies. Small solitary bees (Dufourea, Andrena) visit the flowers in considerable numbers at certain times and places, and when they do make visitations they are highly effective as pollinators.

Gilia cana speciosa, which occurs in a few canyons at the south base of the Sierra Nevada, is distinguished by its purple salverform flowers with a long slender tube 2.0 to 2.5 cm. long. The mode of pollination of these long-tubed plants remains a mystery after years of unsuccessful observations. Butterflies and small beeflies have been seen on the flowers as casual visitors and only partially successful feeders. Other species of flowers with long slender tubes in the California flora are known to be pollinated by the long-tongued fly, Eulonchus (Cyrtidae), but as yet Eulonchus has not been found in the area of *Gilia cana speciosa*.

FLORAL MECHANISM. *Gilia cana* consists of five named subspecies which occur in different areas of the Mojave Desert and possess different floral characters. Most of the races, such as *G. cana triceps* and others, have pink, slender, funnelform flowers with a tube 0.5 to 1.5 cm. long. The interesting endemic race, *G. c. speciosa*, which occurs in several desert canyons at the base of the southern Sierra Nevada, has very fragrant, purple, salverform flowers with a long slender tube 2.0 to 2.5 cm. long.

The style is usually elevated above the short stamens. However, forms of *G. cana triceps* are known in which the stigmas and anthers develop on the same level.

BREEDING SYSTEM. *Gilia cana triceps* is self-compatible, but largely outcrossing, though autogamous individuals are found as rare variants in some polymorphic populations. *Gilia cana speciosa* is also self-compatible and non-autogamous, but here the I_1 generation contains many inviable individuals, and in later inbred generations the general vigor declines until the lines eventually become extinct.

INSECT VISITORS.

AMARGOSA MTS., INYO COUNTY, CALIFORNIA   (*G. c. speciformis*)
Oligodranes cinctura (Bombyliidae).

DAGGETT, SAN BERNARDINO COUNTY   (*G. c. triceps*)
Oligodranes sp. (Bombyliidae).

GOODSPRINGS, CLARK COUNTY, NEVADA   (*G. c. triceps*, David P. Gregory)
Dufourea sp. (Halictidae). *Phthiria sp.* (Bombyliidae). Melyrid beetle (casual).

SAN BERNARDINO MTS., SAN BERNARDINO COUNTY   (*G. c. bernardina*)
Oligodranes sp. (Bombyliidae).

ROCK CREEK, MONO COUNTY   (*G. c. cana*)
Andrena chapmanae (Andrenidae). *Bombylius sp.* (Bombyliidae).

BIG PINE CREEK, INYO COUNTY   (*G. c. cana*)
Bombylius lancifer (Bombyliidae).

SHORT CANYON, KERN COUNTY   (*G. c. speciosa*)
Vanessa carye (Nymphalidae). *Bombylius sp.* (Bombyliidae). Melyrid beetles.

RED ROCK CANYON, KERN COUNTY   (*G. c. speciosa*)
Oligodranes sp. (Bombyliidae). *Villa sp.* (Bombyliidae).

Gilia diegensis

This species of the interior mountains of coastal southern California has small to medium-sized funnelform flowers. The plants are self-compatible and at least partly autogamous. They are also on occasion actively visited and cross-pollinated by small solitary bees, such as Dufourea, Chelostomopsis, and others, which settle in the small corolla throat and carry pollen on various parts of the body.

FLORAL MECHANISM. The flowers are small to medium-sized, funnelform, and moderately colorful with a pinkish-violet limb and yellow throat. In a population at Oak Glen in the San Bernardino Mts. for which pollination records are available, the limb is 7 or 8 mm. broad, the throat about 2 mm. broad, and the throat and tube together 4 mm. long. The short stamens stand just above the throat and the style rises slightly above the anthers.

BREEDING SYSTEM. Three strains from Riverside and San Diego counties have been found to be self-compatible and autogamous. The I_1 progeny are fully vigorous.

INSECT VISITORS. The following bees were observed on the flowers in the Oak Glen population in May 1955: *Andrena sp.* (Andrenidae). Dufourea (2 spp., Halictidae). *Chelostomopsis rubifloris* (Megachilidae). *Anthophora urbana* (Apidae).

Gilia brecciarum

This species ranges from the interior mountains of southern California across the Mojave Desert to Nevada. It is self-compatible and does not suffer from any known inbreeding depression. In one of the races of the Mojave Desert, *G. b. neglecta*, the flowers are medium large, and bright violet, purple, and yellow, with a broad throat and short tube. This race is non-autogamous. Its flowers are pollinated chiefly by small solitary bees (Dufourea) and also by beeflies (*Bombylius lancifer*). Another race, *G. b. brecciarum*, with small dull-colored flowers is autogamous.

FLORAL MECHANISM. The flowers are funnelform, open by day, canted upward and outward in the somewhat congested inflorescence, and contain nectar at the base of the corolla tube. The three named subspecies differ in the size and colorfulness of the flowers.

In *G. b. neglecta* in the Mojave Desert the flowers are medium large, colorful, and have a broad throat and short tube. The floral characteristics of a population in Short Canyon, Kern County, for which pollination records are available, are typical of this subspecies. The corolla is violet in the limb with a purple eye in the throat set off by bright yellow spots. The limb is 1.5 cm. broad, the throat 3 mm. broad and 5 mm. deep, and the tube 5 mm. long. The anthers stand above the throat and the stigma is exserted well beyond the anthers.

In *G. b. argusana* in a different part of the Mojave Desert, the flowers differ from the above in having a longer corolla tube.

In *G. b. brecciarum*, which has a scattered distribution from southern California to Nevada, the flowers are small, dull-colored, and self-pollinating.

BREEDING SYSTEM. Five strains of *G. brecciarum* representing each of the subspecies have been found to be self-compatible. The I_1 generation has been grown for three of these strains, namely the Short Canyon population of *G. b. neglecta* and two populations of *G. b. breccairum*, and found to be fully vigorous. *Gilia brecciarum neglecta* is non-autogamous, while *G. b. brecciarum* is autogamous.

INSECT VISITORS. The Short Canyon population of *G. b. neglecta* was being visited actively by several individuals of *Dufourea sp.* (Halictidae) in the spring of 1962. This small solitary bee was poking its head into the corolla throat and getting nectar out of the tube with its short proboscis (ca. 2 mm. long). It was carrying pollen from flower to flower on its head and venter.

In two other populations of *G. b. neglecta* in the Mojave Desert, *Bombylius lancifer* has been noted as a consistent flower visitor and pollinator.

A melyrid beetle is an occasional visitor, and as it carries a few pollen grains on its back it may play a minor role in the pollination of *G. b. neglecta*.

Gilia latiflora   *Plate I E*

This species of showy large-flowered annuals occurs chiefly in the Mojave Desert and extends into the arid valleys of the inner South Coast Ranges. The flowers are fragrant, funnelform, often with a very ample throat, and colorful with a white or violet limb and yellow and purple markings in the throat. The proportions and coloration of the flowers differ from one geographical race to another.

The breeding system also varies geographically. In the species as a whole the floral mechanism is such as to attract insects and promote outcrossing. The populations in the Mojave Desert are non-autogamous but differ in self-compatibility. A series of populations in the western and northwestern part of the desert are self-compatible, while some populations farther east are self-incompatible. Partial autogamy is found in the plants of the South Coast Ranges.

Two of the subspecies in the Mojave Desert, *G. l. latiflora* and *excellens*, are alike in having large, full-throated, brightly colored flowers, but differ in the length of the corolla tube, which is short in subspecies *latiflora* but long and stout (0.9 to 1.4 cm.) in subspecies *excellens*.

The chief pollinator of the full-throated races mentioned above is the large black bee Tetralonia (Apidae), which has been seen repeatedly on the flowers in different populations and in different years. This bee settles into the broad corolla throat, which accommodates it well, often to collect pollen, and gets much Gilia pollen on its venter and legs in the process. It flies rapidly from flower to flower and plant to plant.

Also effective as pollinators, but less consistent as visitors, are other large bees such as Anthophora, Emphoropsis, and Megachile. The flowers are also visited by various beeflies, which can bring about some pollination.

The adaptive significance of the long stout corolla tube in *G. l. excellens* cannot be stated with certainty. The long tube of this race can scarcely be considered an adaptation to its most effective known pollinator, Tetralonia, which has a short proboscis. On one occasion we saw the hawkmoth, *Celerio lineata*, feeding on and pollinating *G. l. excellens* by day, but we never observed this again in subsequent years.

FLORAL MECHANISM. Gilia latiflora is a polytypic species of the Mojave Desert and the inner South Coast Range valleys. The geographical races, differing in floral as well as vegetative characters, are grouped into six named subspecies. Four of these subspecies have been tested as to breeding system, as will be described below. Our pollination records pertain to two subspecies, namely *G. latiflora latiflora* and *G. l. excellens*, which will be emphasized in the following account of floral mechanisms.

In *Gilia latiflora* as a whole the flowers are medium large, colorful, fragrant, nectariferous, funnelform, and borne on short pedicels. The corolla is violet to white with purple and yellow markings in the throat. The anthers stand at three levels above the corolla throat, and the stigma is raised beyond the anthers.

In the full-throated race of G. l. *latiflora* in the Mojave River drainage area, the corolla limb is 2.0 to 2.5 cm. broad, the throat 6 to 7 mm. wide and 7 to 8 mm. deep, and the tube 5 to 6 mm. long. Farther north in the El Paso Mts. is G. l. *excellens*, which is like G. l. *latiflora* in possessing a broad corolla throat but unlike it in having a long stout tube, 0.9 to 1.4 cm. long.

BREEDING SYSTEM. Seven strains belonging to four subspecies have been tested in the screenhouse in Claremont. All strains of subspecies *latiflora*, *davyi*, and *excellens* are non-autogamous. The Lockwood Valley strain of G. l. *cuyamensis* is partially autogamous in that some untouched caged flowers, but not all, set capsules which were smaller than normal.

Self-compatibility has been found in the Mojave strain of G. l. *latiflora*; the Gorman and Antelope Valley strains of G. l. *davyi*; and the Lockwood Valley strain of G. l. *cuyamensis*. The I_1 generations of the Gorman and Antelope Valley strains contained many vigorous individuals and an occasional inviable segregate.

Self-incompatibility certainly or probably occurs in the Apple Valley and Adelanto strains of G. l. *latiflora* and the Johannesburg strain of G. l. *excellens*. The Apple Valley strain is concluded to be self-incompatible from the fact that 26 selfed flowers set no capsules, whereas a similar number of cross-pollinated flowers set capsules freely. The Johannesburg strain is probably also highly self-incompatible but needs to be retested. The Adelanto strain is only partially self-incompatible, in that 20 self-pollinated flowers produced 16 capsules which were smaller than normal.

INSECT VISITORS.

GILIA LATIFLORA LATIFLORA (3 populations)
Tetralonia californica (Apidae). *Apis mellifera* (occasional). *Bombylius sp.* (Bombyliidae). *Oligodranes sp.* (occasional, Bombyliidae).

GILIA LATIFLORA EXCELLENS (3 populations)
Megachile sp. (Megachilidae). *Anthophora sp.* (Apidae). *Emphoropsis sp.* (Apidae). *Tetralonia sp.* (Apidae) (Alva Day in 1955; R. W. Tharp and C. Epling in 1962; V. G. and K. G. in 1962). *Lordotus albidus* (Bombyliidae). *Oligodranes cinctura* (Bombyliidae). *Aphoebantus sp.* (Bombyliidae). *Celerio lineata* (Sphingidae).

Gilia tenuiflora   Fig. 13

Gilia tenuiflora is a slender plant with violet flowers in the South Coast Ranges of California. The plants are self-compatible and partially autogamous in some populations. The flowers are narrowly funnelform with a slender throat and tube 2 mm. broad at the orifice and 8 mm. long.

The most numerous and effective pollinator in a population near Creston is Oligodranes (Bombyliidae). The body of this small beefly is 3 mm. long and its proboscis is 2 mm. long. It is thus the right size to fit into the slender corolla throat and to probe for nectar. It carries pollen on its body from flower to flower.

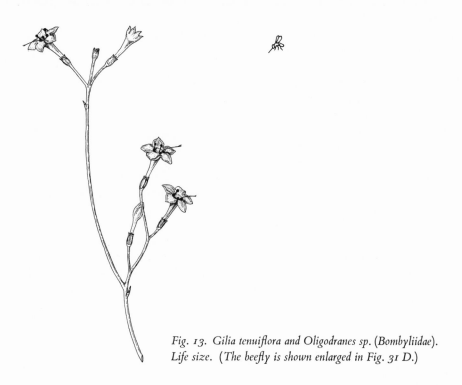

Fig. 13. Gilia tenuiflora and Oligodranes sp. (Bombyliidae).
Life size. (The beefly is shown enlarged in Fig. 31 D.)

FLORAL MECHANISM. The flowers are deep violet with a white ring around the orifice, slender funnelform, canted to the side, and with slightly exserted stamens. In a population near Creston, San Luis Obispo County, the throat measures 2 mm. in diameter and the tube and throat together are 8 mm. long.

BREEDING SYSTEM. A strain from Arroyo Seco, Monterey County, is self-compatible and partially autogamous, in that 15 out of 38 flowers set capsules autogamously. The I_1 progeny are fully vigorous.

INSECT VISITORS. At Creston, we found *Oligodranes sp.* (Bombyliidae) visiting and pollinating the flowers. A melyrid beetle was also present as a casual visitor.

Gilia leptantha   Plates I F, III J

Gilia leptantha grows in open sandy places in the pine belt of several mountain ranges in southern California. Three of the four disjunct geographical races have been studied from the standpoint of breeding system and flower pollination. These races are *G. l. pinetorum* on Mt. Pinos, Ventura County; *G. l. purpusii* in the southern Sierra Nevada; and G. *l. leptantha* in the San Bernardino Mts. farther south.

Gilia leptantha pinetorum has blue-violet and yellow funnelform flowers with a small throat and short tube and with exserted stamens and style. It is self-compatible but non-autogamous. The flowers are visited and pollinated by various bees, most effectively by small solitary bees such as Lasioglossum which fit well into the corolla throat.

The other two geographical races, *purpusii* and *leptantha*, have subsalverform flowers with long slender corolla tubes 0.7 to 2.2 cm. long. The long slender tubes suggest visitations by long-tongued flies like Bombylius and Eulonchus, which are known to feed on and pollinate other flowers of similar size and shape in the Phlox family, but these insects have not yet been observed on G. *l. purpusii* or *leptantha*.

A population of G. *l. leptantha* in the San Bernardino Mts. is polymorphic for flower size and breeding system. Some individuals have long corolla tubes (up to 1.8 cm. long) and long styles, and breed true for this condition in garden-grown progeny. Other true-breeding individuals have short tubes (0.8 cm.) and short styles. Intermediate forms are also present. All individuals are self-compatible. But the long-tubed long-styled individuals are non-autogamous, while the short-tubed short-styled individuals are autogamous.

Originally polymorphic experimental populations when isolated from insects quickly become monomorphic for the autogamous condition. The natural population has remained polymorphic during at least several generations of observation. The short-tubed autogamous biotypes are perpetuated in this population, not only by the self-pollinating mechanism, but also by preferential flower visitations by relatively short-tongued beeflies (Villa and Oligodranes), which can reach the nectar in these short flowers. The factors responsible for the perpetuation of the long-tubed non-autogamous biotypes are not known but are plausibly suggested to be long-tongued flies.

FLORAL MECHANISM. *Gilia leptantha*, a polytypic species in the pine belt of southern California and the southern Sierra Nevada, consists of four main geographical races, three of which have been studied from the standpoint of pollination. The three races in question are: (1) G. *l. pinetorum* on Mt. Pinos, Ventura County; (2) G. *l. purpusii* on the Kern River drainage in the southern Sierras; and (3) G. *l. leptantha* in the San Bernardino Mts.

These races are alike in having long exserted stamens and still further exserted styles. They differ strikingly in the length of the corolla tube. *Gilia leptantha pinetorum* has short-tubed funnelform flowers, and G. *l. purpusii* and *leptantha* have long-tubed subsalverform flowers.

The flowers of G. *l. pinetorum* are blue-violet with a yellow throat. They are funnelform, as stated above, with a throat 2 or 3 mm. wide which passes into a tube 4 to 7 mm. long.

In G. *l. purpusii* the corolla has a bright pinkish-violet limb and purple (or yellow) tube. The corolla is subsalverform, with a throat which varies from small to moderately full, and a tube which varies in length. In many populations the tube is 1.1 or 1.2 cm. long; it reaches the extreme length of 2.2 cm. in some populations; and may vary considerably within a single polymorphic population, as at Johnsondale, where different individuals have tubes ranging from 0.7 to 1.4 cm. long.

The flowers of G. *l. leptantha* are also subsalverform and variable in tube length. The corolla limb is bright pink and the tube pure yellow or yellow tinged with purple. The tube varies from 0.8 to 1.8 cm. long in different individuals in the same polymorphic population. In some short-tubed individuals the stigma is on the same level as the anthers.

BREEDING SYSTEM. *Gilia leptantha pinetorum* from Mt. Pinos is self-compatible but non-autogamous. Three populations of G. *l. purpusii* on the Kern River are non-autogamous and probably but not certainly self-compatible.

A population of G. *l. leptantha* on the upper Santa Ana River in the San Bernardino Mts. is interesting in being polymorphic for both the flower size, as noted in a preceding paragraph, and for breeding system. Several long-tubed and long-styled individuals tested in the screenhouse are self-compatible but non-autogamous, while the short-tubed short-styled plants set seeds autogamously. In two generations of cultivation in the insect-free screenhouse the polymorphism was lost, as the autogamous types completely supplanted the non-autogamous forms. In the natural population the polymorphism has been maintained over a period of observation of several years and no doubt for a longer time.

INSECT VISITORS.

GILIA LEPTANTHA PINETORUM   (Mt. Pinos)
  Lasioglossum trizonatum (Halictidae). *Chloralictus sp.* (Halictidae). *Anthophora urbana* (Apidae). *Apis mellifera* (Apidae). *Bombylius sp.* (Bombyliidae).

GILIA LEPTANTHA PURPUSII   (Johnsondale)
  Evylaeus sp. (Halictidae). *Agapostemon cockerelli* (Halictidae). *Oligodranes sp.* (Bombyliidae). *Trichochrous suturalis* (Melyridae).

GILIA LEPTANTHA LEPTANTHA   (San Bernardino Mts.)
  Villa alternata group (Bombyliidae). *Oligodranes sp.* (Bombyliidae).

The autogamous species   *Fig.* 14

The majority of species in the section Arachnion have small dull-colored flowers which set seeds autogamously and produce fully vigorous I_1 progeny.

This breeding system has been demonstrated experimentally for the following species: G. *austrooccidentalis, clokeyi, crassifolia, inconspicua, interior, jacens, malior,*

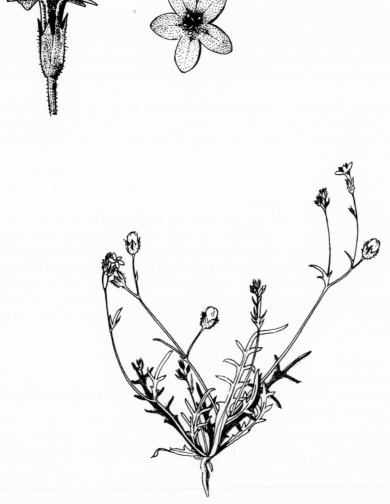

Fig. 14. Gilia minor. Plant life size; flowers × 3.

mexicana, minor, modocensis, ophthalmoides, sinuata, tetrabreccia, transmontana, and tweedyi.

Autogamy is also developed in particular races of some other species, thus in G. aliquanta breviloba, G. brecciarum brecciarum, G. diegensis (unnamed local races), G. flavocincta australis, and G. ochroleuca ochroleuca.

Many of the autogamous species and subspecies attain a widespread distribution in the western American deserts. *Gilia crassifolia* occurs in Patagonia and Chile.

Insect visitations to the autogamous flowers are uncommon but do occur occasionally. For example, we have records of Dufourea visiting G. *malior* in the sagebrush plains near Litchfield, Lassen County, California, and of Oligodranes visiting G. *sinuata* in Fish Lake Valley, Esmeralda County, Nevada. However, in a good spring in the desert the autogamous Cobwebby Gilias are present in a local area in countless millions of individuals, whereas their potential pollinating insects are few and far between, and under these conditions the great majority of plants must reproduce autogamously.

WOODLAND GILIAS

The section Saltugilia, or Woodland Gilias, contains eight species of large or medium-sized annuals of western North America. The cross-fertilizing and morphologically more primitive species occur in openings in pine forest on the Pacific slope. Several reduced and autogamous species are found in high mountains and on the western edge of the desert.

Gilia caruifolia

Gilia caruifolia is a large scapose herb in the mountains of San Diego County, California, and northern Baja California. Its pale blue-violet funnelform flowers are borne loosely at the ends of the branches. The corolla throat is 3 mm. wide, the nectar-containing tube is short, and the stamens and style are elevated above the throat. A population on Palomar Mt., San Diego County, is self-incompatible. The flowers are visited and cross-pollinated mainly by small solitary bees, particularly *Halictus tripartitus* in one period of observation, which enter the corolla throat and pick up and carry pollen on their venter.

FLORAL MECHANISM. The flowers are funnelform and pale blue-violet with yellow and often also purple spots in the upper throat. The corolla limb is 1.5 to 2.0 cm. broad, the throat 3 mm. broad by 3 mm. deep, and the tube 3 to 5 mm. long. The stamens are well exserted at different levels and on the lower side of the corolla in its tilted position. The style also lies on the lower side of the corolla and extends 2 or 3 mm. beyond the anthers. The flowers are protandrous with an overlap in the staminate and pistillate phases (V. and A. Grant, 1954).

BREEDING SYSTEM. A strain from Palomar Mt., San Diego County, was caged and tested.

The 15 self-pollinated flowers were completely sterile, whereas the sib crosses were fully fertile, indicating that this strain is self-incompatible (V and A. Grant, 1954).

INSECT VISITORS. The main pollinator of the Palomar Mt. population during one period of observation was the small solitary bee, *Halictus tripartitus* (Halictidae). A small beetle, *Listrus famelicus* (Melyridae), was present in the flowers as an occasional visitor and relatively ineffective pollinator. This beetle is a common visitor and effective pollinator of Calochortus in the same region. Another occasional visitor and non-pollinator was the small moth, Adela (Adelidae) (V. and A. Grant, 1954).

Gilia splendens   *Plates II G, III H; Fig.* 15

Gilia splendens, a species closely related to *G. caruifolia*, grows in pine woods in the South Coast Ranges and the San Gabriel, San Bernardino, and San Jacinto Mts. of California. The flowers are pink and funnelform with exserted stamens and styles. The plants are self-compatible and usually non-autogamous.

Four geographical races differ in flower form and mode of pollination. (1) The widespread race has funnelform flowers with a short tube (5 to 6 mm.). (2) In the San Gabriel Mt. race the flowers have a long slender tube ($\pm$ 1.5 cm.). (3) A local race in the San Bernardino Mts. has intense pink flowers with a long stout tube and throat (1.5 to 2.0 cm.). (4) A desert race is characterized by pale-colored flowers with a short tube and short style.

The widespread race is visited and pollinated chiefly by *Bombylius lancifer*, on the basis of observations in seven populations in southern California (V. and A. Grant, 1954). This beefly enters the slender corolla throat partway, while hovering, and inserts its proboscis into the corolla throat for nectar. The proboscis, which varies from 5 to 10 mm. long in different individuals, corresponds in length to the slender lower throat and tube. In probing for nectar, the beefly picks up and carries pollen on its venter and contacts the stigma with the same body part. Some supplementary pollination of this race is carried out by pollen-collecting halictid bees.

As regards the San Gabriel Mt. race, the adaptive role of its very long slender tube was not understood for some years despite considerable field study. *Bombylius lancifer* visits the flowers of this race frequently, but, owing to its relatively short tongue, cannot reach the main body of nectar at the base of the long tube. Pollen-collecting halictid bees visit the flowers occasionally. Both the beeflies and bees do bring about pollination. But the distinctive racial character of the flowers, the long slender tube, can scarcely be explained as an adaptation to these relatively short-tongued insects.

A

B

C

D

Hummingbird bills, though long enough, are too broad to fit inside the slender corolla tube, as the senior author has determined by trying to insert the bills of dead specimens into live flowers. In nature where the San Gabriel race of *G. splendens* is growing with species of Penstemon possessing red trumpet-shaped flowers, the hummingbirds pay no attention to the Gilia; but where the latter is present alone the birds sometimes feed on it, apparently by exserting their extensile tongues down the tube. Hummingbird visits of this sort can bring about pollination.

The critical observation of a flower-feeding animal well coadapted with the San Gabriel race of *Gilia splendens* was finally made on a sunny spring morning in 1961. A population of the Gilia was being visited by a number of individuals of the metallic golden-blue fly, *Eulonchus smaragdinus* (Cyrtidae), which has a very long slender proboscis. The needle-like bill of this fly, which measures 1.1 to 1.4 cm. long in most specimens in this area, fits perfectly down the long slender corolla tube ($\pm$ 1.5 cm.). The fly approaches the flowers with its proboscis extended forward, hovers or settles lightly with its proboscis inserted, and picks up pollen on its venter in the process. The flies go actively and systematically from flower to flower in the Gilia population.

Eulonchus is fairly secretive in its habits, a rapid and wary flier, and is easy to overlook in an area. It is stated to be "uncommon" and "rare" in the entomological literature. We were naturally hesitant at first to conclude that the San Gabriel race could be specialized for pollination by a rare and perhaps unreliable agent. But since learning when and how to look for Eulonchus, we have since found it on repeated occasions on different populations of the San Gabriel race. It is a regular visitor of these flowers.

The third race occurs in the upper pine zone in a localized area in the San Bernardino Mts. It is distinguished by brilliant pink and nearly trumpet-shaped flowers with a broad long tube and throat (up to 2 cm.). Near Strawberry Peak in July 1955, Dr. Richard Beeks and the senior author saw female hummingbirds feeding on the flowers in their usual hovering position with the bill inserted into the flower tube. In this position the base of the bill contacts the exserted stamens

Fig. 15. The races of Gilia splendens and their normal pollinators. All life size.

(A) Widespread race and Bombylius lancifer (Bombyliidae).
(B) High San Gabriel Mt. race and Eulonchus smaragdinus (Cyrtidae).
(C) High San Bernardino Mt. race and Stellula calliope (Trochilidae).
(D) Desert race, largely autogamous.

and style. The birds were flying systematically from plant to plant in the Gilia population, occasionally visiting the red tubular flowers of a neighboring Penstemon, but feeding mainly on the Gilia.

The bills of three common species of hummingbird in the area have the following lengths measured from tip of bill to base of feathered muzzle: *Calypte anna* (1.8 to 2.0 cm.); *Selasophorus rufus* (1.7 to 2.0 cm.); *Stellula calliope* (1.5 to 1.7 cm.). The extensile tongue can probe well beyond the bill tip. It will be noted that the San Bernardino race of flowers corresponds well in length with the bills of the local hummingbirds. The bills of dead hummingbird specimens fit easily into live flowers and pollen remains adhering to the specimen after a simulated probe.

The San Bernardino race is also visited by *Bombylius lancifer*, halictid bees, and honeybees. The beeflies cannot reach the main nectar supply but apparently obtain some capillary nectar in the upper tube, and the bees collect pollen. These insects bring about supplementary pollination.

The desert race with its smaller paler flowers and shorter styles sets seeds autogamously in the absence of insect visits and is cross-pollinated occasionally by small solitary bees and no doubt also by other insects.

The differences between the races of *Gilia splendens* in mode of pollination are relative rather than absolute. The total range of animal pollinators is known to be similar for the different races of this species. *Bombylius lancifer* visits and pollinates the flowers of both the widespread and the San Gabriel races; similarly, hummingbirds feed on both the San Gabriel and San Bernardino races. Yet the races correspond morphologically to the proboscis or bill proportions of different members of their total range of animal pollinators. The San Bernardino race is specialized primarily for hummingbird pollination, though it is not pollinated exclusively by hummingbirds; while conversely, the San Gabriel race is pollinated to some extent by hummingbirds, yet is specialized for a different agent, Eulonchus.

The racial differences in floral characters can be explained on the grounds that the relative frequency of effective pollinating visits by different species of flower-visiting animals changes from one geographical area to another within the range of *Gilia splendens*. The observational evidence, so far as it goes, supports this premise. Hummingbirds are only occasional visitors and relatively ineffective pollinators of the San Gabriel race, but are regular and effective pollinators of the San Bernardino race, which is specialized for bird pollination. The San Gabriel race, on the other hand, is specialized for its regular and effective Eulonchus pollinators and not for its occasional bird visitors.

FLORAL MECHANISM. Gilia splendens grows in openings in pine woods in the coastal mountains of California from the South Coast Ranges to the San Jacinto Mts. Throughout this area, four geographical races may be recognized, which differ in floral characters and pollination system. These are: (1) a widespread race in the South Coast Ranges, San Jacinto Mts., and elsewhere; (2) a race in the higher parts of the San Gabriel Mts.; (3) a race occupying a local area in the high San Bernardino Mts.; and (4) a race on the western edge of the desert just east of the San Bernardino and San Jacinto Mts.

The flowers in the species as a whole are slender funnelform and pink with exserted blue anthers and a still further exserted style. The flowers lie canted to the side with the lowermost corolla lobes pointed forward, the upper one slightly reflexed, and the style exserted on the lower side. The stamens ripen before the stigma.

The distinctive features of the various races are as follows: (1) Widespread race: tube short (5 to 6 mm. long); tube and throat about 1.2 cm. long. (2) San Gabriel race: tube long and slender (often 1.5 to 1.8 cm.); tube and slender throat 1.8 to 2.3 cm. long. A complete series of intergrades in tube length connect this with the preceding race. (3) San Bernardino race: flowers intense pink; tube and throat long and stout (1.5 to 2.0 cm.). (4) Desert race: flowers pale; tube short (3 mm.); style shorter than in the preceding races.

BREEDING SYSTEM. Gilia splendens is self-compatible on the basis of tests with strains of the San Gabriel and desert races. Most populations are non-autogamous, but the desert race sets seeds autogamously (V. and A. Grant, 1954).

ANIMAL VISITORS.

WIDESPREAD RACE (7 populations)
   Lasioglossum sisymbrii (Halictidae). *Dufourea sp.* (Halictidae). *Bombylius lancifer* (regular visitor, Bombyliidae). *Syrphus americanus* (Syrphidae). *Anthaxia aenogaster* (probably not pollinating, Buprestidae).

SAN GABRIEL RACE (4 populations)
   Lasioglossum sisymbrii (Halictidae). *Bombylius lancifer* (Bombyliidae). *Aphoebantus sp.* (Bombyliidae). *Eulonchus smaragdinus* (Cyrtidae). Unidentified syrphid fly. *Anthaxia aenogaster* (Buprestidae). Unidentified female hummingbird.

SAN BERNARDINO RACE (1 population)
   Halictus farinosus (Halictidae). Apis mellifera (Apidae). *Bombylius lancifer* (Bombyliidae). Unidentified syrphid fly (eating pollen). Unidentified female hummingbird.

DESERT RACE (1 population)
   Unidentified small solitary bees.

Gilia leptalea

Gilia leptalea is a common species of small annual plant in the pine belt of the Sierra Nevada and North Coast Ranges. The bright pink, slender funnelform flowers are similar in shape but smaller in size than those of *G. splendens*. A Sierran strain is self-compatible. The chief agents of pollination in two Sierran populations are small bees and small beeflies, which can and do enter the slender corolla throat.

FLORAL MECHANISM. The flowers are faintly sweet-scented, bright pink, and funnelform with a long tapering throat and slender tube. The corolla throat and tube together measure about 1 cm. in length; the throat is 2 mm. wide at the orifice. The anthers stand around the throat orifice and the style is exserted.

BREEDING SYSTEM. A strain from Mather, Tuolumne County, proved to be self-compatible (V. and A. Grant, 1954).

INSECT VISITORS. Two populations in the Sierra Nevada, one near Kyburz, El Dorado County, and the other in Ebbett Pass, Calaveras County, have been observed for pollinators (V. and A. Grant, 1954). *Dufourea versatilis* (Halictidae). *Dianthidium ulkei* (Megachilidae). *Lepidanthrax inaurata* (Bombyliidae). *Oligodranes setosus* (Bombyliidae).

The autogamous species

Gilia australis, a relative of *G. splendens*, occurs in the brushy foothills of coastal southern California and onto the Mojave Desert. Its flowers are small and autogamous, and the I_1 progeny are fully vigorous (V. and A. Grant, 1954).

Gilia stellata, which attains a wider distribution in the deserts of California and Arizona, is also autogamous.

Gilia capillaris, a reduced and small-flowered relative of *G. leptalea*, occurs at high elevations in the mountains of the Pacific slope with disjunct stations in the Rocky Mts. It too is autogamous. We have observed *Bombylius lancifer* and melyrid beetles visiting the flowers of a population in the San Bernardino Mts. The beeflies were standing off too far from the small flowers to contact the anthers and stigma. The beetles, however, had a few pollen grains on their bodies and could function as rather ineffectual pollinators.

6

IPOMOPSIS

Ipomopsis is a genus of summer-blooming herbaceous plants related to Gilia. The approximately 27 species fall naturally into three sections. The main center of distribution of the genus lies in the Rocky Mt. region and the Southwest. A number of species also occur on the Pacific slope and in northern Mexico, while one species is found in the southeastern United States, and another in southern South America. The flowers are tubular or salverform in shape. In the first two sections of the genus they are usually large and colorful—often bright red, violet, or white; and in the third section they are small, whitish, and often clustered in heads.

SECTION PHLOGANTHEA

This section consists of six species of spreading perennials, often woody-based, in the Southwest and Mexico.

Ipomopsis multiflora　　　　　　　　　　　　　*Plate I K*

This spreading perennial plant occurs in the mountains of Arizona and New Mexico. The violet salverform flowers are grouped in small heads. The slender sinuous corolla tube is curved upward at the base and downward near the orifice, and is about 7 mm. long. The stamens and style are exserted beyond the orifice, the former to a distance of 4 mm.

In the Santa Rita Mts. of Arizona we have observed bumblebees, *Bombus sonorus*, visiting the flowers and effecting venter pollination. Further observations would undoubtedly extend the list of bee visitors.

The related species, *I. pinnata* in Mexico and *I. polyantha* in the Southwest, have similar floral characteristics suggestive of bee pollination.

Ipomopsis tenuifolia Plate II D

Ipomopsis tenuifolia is a woody-based perennial endemic in the arid mountains of a small region in southern San Diego County, California, and adjacent Baja California. The plants are self-incompatible. The deep red flowers are trumpet-shaped with a broad tube 1.5 cm. long and exserted stamens and style. The animals best fitted to feed on and pollinate these flowers are hummingbirds.

FLORAL MECHANISM. The flowers are deep reddish-violet with white markings on the lobes around the orifice and are trumpet-shaped. The tube is 1.5 cm. long and 3 mm. broad at the orifice and narrows down in the nectar-containing base. The stamens are exserted 1 cm. beyond the orifice and the stigma several millimeters beyond the anthers. The flowers are strongly protandrous.

BREEDING SYSTEM. A strain from Jacumba, San Diego County, California, is self-incompatible.

ANIMAL VISITORS. Plants grown in the Rancho Santa Ana Botanic Garden are commonly visited by hummingbirds. Their bills slip easily into the tube, as they hover and probe for nectar and their heads become dusted with pollen at the same time. Hummingbirds are probably the principal pollinators of *I. tenuifolia* also in nature.

SECTION IPOMOPSIS

This section contains nine species of tall herbaceous plants, usually biennial but sometimes annual or perennial, which range from the Rocky Mt. region west to the Pacific slope and east to Florida. The flowers are large, tubular, and colorful in all members of the section. Beyond these common characteristics, a division is evident between the species possessing red trumpet-shaped flowers pollinated by hummingbirds and those with violet or white long-tubed salverform flowers pollinated by hawkmoths.

Ipomopsis rubra

This tall biennial herb grows in open sandy places from Texas to North Carolina and Florida. Its bright red, trumpet-shaped flowers, borne in showy panicles, are visited and pollinated by Ruby-throated hummingbirds (*Archilochus colubris*).

FLORAL MECHANISM. The plants bloom from late May or June through the summer. The flowers are borne on short pedicels in loose or dense panicles at the tops of the erect stems. They are bright scarlet outside and yellow streaked with red inside. The trumpet-shaped corolla tube is 2 cm. long and 5 to 7 mm. wide at the orifice, narrowing down to the nectar-containing base. The corolla lobes, yellow stamens, and style extend about 1 cm. beyond the orifice (Wherry, 1936).

ANIMAL VISITORS. Ruby-throated hummingbirds (*Archilochus colubris*) are common visitors of the flowers (Wherry, 1936). The bills of these birds, measuring 1.7 to 2.0 cm. long from tip to base of feathered muzzle, correspond well with the corolla tube length.

Ipomopsis arizonica

This low herbaceous plant inhabits the piñon zone in the desert mountains of eastern California, southern Nevada, and northern Arizona. The senior author formerly regarded it as a geographical race of the more widespread *I. aggregata*, along with Fosberg and other authors (Fosberg, 1942; Grant, 1956), whereas Wherry (1946, 1961) has consistently treated it as a separate species, and we are now inclined to follow Wherry. *Ipomopsis arizonica* and *I. aggregata* are sympatric and non-interbreeding in some localities in the Southwest, as on the north rim of the Grand Canyon, Arizona, where the former occurs in the piñon pine zone and the latter in yellow pines 100 yards away.

The flowers are pure bright red and trumpet-shaped with a relatively short tube (1.2 to 1.5 cm. or rarely longer). The stamens are included. Hummingbirds (unidentified) have been seen visiting and pollinating the flowers of a population in the Charleston Mts. of Nevada.

Ipomosis aggregata   Plate II E, Fig. 16

This tall biennial herb with, usually, red trumpet flowers occurs over a very large area from the Rocky Mts. to the Sierra Nevada and Cascade Mts., and in a wide variety of habitats from semi-arid pine-oak woods to high mountain meadows and sagebrush scrub. As would be expected, the species consists of numerous geographical and ecological races. The populations comprising some of the races are, moreover, highly polymorphic, whereas those constituting other races are uniform. The geographical and polymorphic variation includes floral as well as vegetative characters.

Ipomopsis aggregata hybridizes with two species of long-tubed whitish hawk-moth flowers, *Ipomopsis candida* and *I. tenuituba*, in the eastern and western parts of its range respectively. There are localities where the two species (or semispecies) grow together sympatrically without hybridizing, but in other localities hybrid

Fig. 16. *Ipomopsis aggregata* (*western race*) *and Selasphorus rufus* (*Trochilidae*). *Life size.*

swarms are found, and in still other places intermediate races of evident hybrid origin have become established, either locally or over a wide area.

The races on the Pacific slope, in the Southwest, and at lower elevations in the southern Rocky Mts. are in general characterized by odorless, red, trumpet-shaped flowers borne in loose racemes, with stamens more or less exserted, and with relatively short and broad corolla tubes. At higher elevations in the Rockies and through the northern Rocky Mt. area the flowers are either odorless or fragrant, generally pink, clustered in denser inflorescences, and have long slender corolla tubes and included stamens.

The principal pollinators of both types of flowers, the red trumpet flowers and the long and slender-tubed pink flowers, are hummingbirds, which feed systematically and regularly in the populations. As the birds probe for nectar in the flower tubes they pick up and carry pollen on their head or bill.

The odorless red-flowered races and the fragrant pink-flowered races differ in their relation to moths. Nocturnal moths have been observed feeding on one population of the latter, and probably bring about considerable pollination of the slender-tubed flowers, though more observations are needed. On the other hand, moths are only occasional feeders and ineffectual pollinators of the trumpet-flowered races.

Some supplementary pollination in all races results from occasional visits by pollen-collecting bees and other insects.

FLORAL MECHANISM. The main racial groupings in *Ipomopsis aggregata* of interest to us from the standpoint of flower pollination are the following:

(1) Pacific slope race. In the yellow pine or sagebrush zone in the northern Sierra Nevada and Cascade Mts. Flowers in loose racemes. Flowers odorless, bright orange-red with yellow or white speckles. Corolla tube relatively short and broad (1.5 to 2.0 cm. long, and 3 to 4 mm. wide at orifice). Stamens well exserted.

(2) Rocky Mt. foothill race (*formosissima, texana*). In the pine and oak zone in the Rocky Mt. region and the Southwest. Flowers similar to the Pacific slope race, but corolla tube longer and more slender (2 to 3 cm. long, 2 to 3 mm. wide at orifice). Stamens slightly exserted to included.

(3) Southern Sierra race (*I. aggregata bridgesii*). In coniferous forest in the southern Sierra Nevada. Flowers magenta. Corolla tube moderately long (2.0 to 2.5 cm.). Stamens included.

(4) Rocky Mt. meadow race (*I. a. aggregata, I. a. attenuata*). In meadows and moist openings in the spruce and aspen zone in the Rocky Mts. Flowers clustered. Flowers fragrant or odorless; generally pink, sometimes salmon or white. Corolla tube long and fairly slender (often 2.5 to 3.0 cm., sometimes less or more). Stamens included in tube. Populations often polymorphic for flower odor, color, and proportions.

(5) Northern Colorado race (extreme form of *I. a. attenuata*). Local at high elevations in the

northern Colorado Rockies. Flowers clustered. Flowers fragrant, cream to pink. Corolla tube very slender and moderately short (± 2 cm.); corolla limb small.

BREEDING SYSTEM. A strain of the Pacific slope race from Plumas County, California, grown in Claremont, was sterile on selfing. The plants were not fully vigorous in the Claremont environment, however, so that physiological inhibitions of fruit formation cannot be ruled out.

Gray (1870, 275) and Darwin (1877) discussed the possibility of heterostyly in *Ipomopsis aggregata* ("*Gilia pulchella*") on the basis of a sampling of dried specimens from different localities. Geographical variation and polymorphism in flower length are of course the features revealed by this comparison, not true heterostyly, which is still unknown in the group.

ANIMAL VISITORS. The regular visitor and pollinator of *Ipomopsis aggregata* are humming-birds, which have been seen on five populations in the Sierra Nevada and six in the Colorado Rockies, and on representatives of all races except the northern Colorado race. Ferguson (1921) also records hummingbirds on a population of the southern Sierra race. Rufous and Calliope hummingbirds (*Selasphorus rufus, Stellula calliope*) have been observed feeding on *Ipomopsis aggregata* in the Sierras; and Broad-tailed hummingbirds (*Selasphorus platycercus*) and less commonly Rufous hummingbirds on the plants in Colorado.

Celerio lineata (Sphingidae) occasionally feeds on the Pacific slope race without pollinating it. Unidentified small moths have been found feeding at night on the flowers of the Rocky Mt. meadow race near Fairplay, Colorado. These flowers could be pollinated by moths.

Pollen-collecting bees visit the flowers fairly commonly. Timberlake (1954) records *Perdita giliae* (Andrenidae) on the flowers of a population near Prescott, Arizona, while we have found *Chloralictus* on flowers at Echo Lake, Sierra Nevada, and bumblebees on a population in Grand Teton National Park, Wyoming.

The beefly *Oligodranes* has been seen on two populations of the southern Sierra race. *Baccha sp.* (Syrphidae) was found eating the pollen of a population of the Pacific slope race.

Ipomopsis tenuituba   *Fig.* 17

This tall herbaceous plant is found at high elevations, often in the aspen zone, in the mountains of the Great Basin and Colorado Plateau. The long-tubed flowers are ascending to horizontal on the branches, have a sweet fragrance, and are pale violet to darker violet.

The long, slender, slightly curved floral tube varies from 3.0 to 4.5 cm. long in different races, and measures 2 or 3 mm. wide at the orifice. At least some of the stamens are included in the tube in all races, and in many races all stamens are included, while the stigma is slightly exserted.

The hawkmoth *Celerio lineata* has been seen feeding on and pollinating the flowers in two populations in Arizona, one on the Kaibab Plateau and the other on Mt. Lemmon. The moths have been seen feeding by day and no doubt also visit the flowers by night. They hover before the flowers and insert their

Fig. 17. Ipomopsis tenuituba and Celerio lineata (Sphingidae). Life size.

proboscis, which varies from 3.2 to 4.5 cm. long in different individuals, down the long corolla tube for nectar. Pollen is picked up and carried on the proboscis.

Some supplementary pollination is carried out occasionally by small pollen-collecting bees, such as Chloralictus in the Kaibab population. The slender flowers and flowering branches bend down under the weight of larger bees and are usually avoided by them.

Ipomopsis candida

Ipomopsis candida occurs in the eastern foothills of the Rocky Mts. and on the adjacent Great Plains. The tall stems bear fragrant, white, long-tubed flowers. The long slender corolla tubes measured 3.2 to 3.6 cm. long in one population on the grassy plains in Colorado and 3.5 to 4.0 cm. long in another mountain population.

The flowers are chiefly pollinated by nocturnal hawkmoths. In the plains population, which was spread out for miles in the grassland, *Celerio lineata* began to feed at dusk and increased its activity after dark. The moths hovered and probed for nectar in the usual manner and went systematically from plant to plant. Two individuals were caught; each had a proboscis 4 cm. long with Ipomopsis pollen on the distal part. A species of Sphinx (near *S. drupiferarum*), with a longer proboscis 4.5 cm. long, was pollinating the mountain population of *I. candida* in the same manner.

Small pollen-collecting bees, such as Chloralictus, visit the flowers by day and bring about some pollination.

FLORAL MECHANISM. The flowers are borne in loose racemes, pointing outward and downward, and are fragrant, white, long-tubed, and slender. Populations may consist wholly or predominantly of white-flowered individuals with admixtures of pink-flowered plants.

The corolla tube is 3.2 to 3.6 cm. long in a population on the grass plains near Port-of Entry, El Paso County; and 3.5 to 4.0 cm. long in a mountain population in Boulder Canyon, Boulder County, Colorado. The spreading corolla lobes form a limb 2 cm. or more broad. The orifice of the tube is 3 mm. wide and the tube narrows down to the nectar-containing base. The anthers are included in the tube 2 to 8 mm. below the orifice. The stigma is exserted slightly. The flowers are protandrous.

INSECT VISITORS. We have observed an extensive plains population at Port-of-Entry and a mountain colony in Boulder Canyon, Colorado. *Celerio lineata* (Sphingidae) was found in the first population; and *Sphinx* aff. *drupiferarum* (Sphingidae) and *Chloralictus sp.* (Halictidae) in the latter.

Ipomopsis thurberi

Our third species of hawkmoth flower in the *Ipomopsis tenuituba* group is *I. thurberi* in the lower elevations of the desert mountains of southern Arizona and northern Mexico. As in the related species considered previously, its flowers are violet with a long slender tube. The floral tube is 3.5 to 3.8 cm. long in most populations, and the stamens are included in this tube. The flowers are visited at night by hawkmoths, particularly *Celerio lineata* in one population studied, which fly systematically and rapidly from plant to plant, carrying pollen on the proboscis.

FLORAL MECHANISM. The flowers are violet with a long, slender, slightly curved tube. The tube measures 3.8 cm. long in a population in the Mule Mts., Cochise County, Arizona, and elsewhere, or may be as short as 3.0 cm. in some local races. When the flowers are freshly opened, the corolla lobes spread out to form a limb 2.5 cm. broad, but in age the lobes bend backward. Nectar may be present in small amounts at the base of the tube by day but its flow increases at dusk. Some of the anthers are slightly exserted, others included in the tube, and the stigma is exserted slightly. The flowers are strongly protandrous.

INSECT VISITORS. We watched a population in the Mule Mts. by day and by night. No visitors were seen during the day. At dusk *Celerio lineata* (Sphingidae) became active in considerable numbers. The proboscis was 3.6 and 3.7 cm. long in two individuals captured and is thus well suited to reach the nectar in these flowers. Pollen adheres to the proboscis.

A large bumblebee, *Bombus sonorus*, was once observed trying to collect pollen from *Ipomopsis thurberi* in the Santa Rita Mts. of Pima County, Arizona. The flowering branch bent down precariously under its weight, and the bumblebee immediately gave up its attempt and did not return to the Ipomopsis flowers during the remainder of the day.

Ipomopsis longiflora   Plate III C

This species is an annual herb which is common in sandy places in the southwestern desert. A strain from the Rio Grande valley is self-compatible. The flowers are pale violet and salverform with a long slender tube. In a population on a desert flat in western New Mexico, this tube ranges from 4.0 to 4.7 cm. long. This population is visited at night by *Celerio lineata*, which probes for nectar from a hovering position, and carries pollen from plant to plant on its proboscis.

FLORAL MECHANISM. The pale violet flowers are slender and salverform with a long gently curved tube and spreading limb. In a typical population near Cienega Lake, Hidalgo County, New Mexico, the tube is 4.0 to 4.7 cm. long and the limb 2.0 to 2.5 cm. broad. Nectar is present at the base of the tube at night. Three of the anthers are exserted slightly, two included in the tube, and the stigma is exserted.

BREEDING SYSTEM. A strain from the Rio Grande valley, Socorro County, New Mexico,

Fig. 18. Ipomopsis macombii and Ochlodes snowi (Hesperiidae). Life size.

grown in Claremont from seeds collected by Dr. David Dunn, proved to be self-compatible. The I_1 generation was fully vigorous.

INSECT VISITORS. We found numerous individuals of *Celerio lineata* visiting the flowers of the Cienega Lake population by night in August 1961. Perhaps 15 or 20 moths were seen in different parts of the large population, all feeding on the nectar and flying actively from plant to plant. Several individuals captured and examined had Ipomopsis pollen adhering to the proboscis or proboscis base.

Ipomopsis·macombii   *Fig.* 18

Ipomopsis macombii is a small-flowered relative of *I. thurberi* in the mountains of the Southwest. Its violet salverform flowers have a slender tube ±1.5 cm. long, which is curved downward, with anthers and stigma standing at or within the orifice.

The most frequent and effective pollinators in a population in the Chiricahua Mts. of Arizona are skippers, *Ochlodes snowi*, which cling to the flowers upside down and insert their proboscis into the tube for nectar. The proboscis, 1.7 to 1.9 cm. long, is adequate to reach the nectar, and picks up pollen in doing so.

Supplementary pollination is carried out by butterflies (*Papilio philenor*) and pollen-collecting bees (Perdita, Anthidium, etc.).

FLORAL MECHANISM. The violet flowers are arranged in one-sided racemes with the corolla tube curved in the middle so the orifice points down. The corolla tube is 1.5 cm. long in a population in the Chiricahua Mts. of Arizona and the orifice is a little more than 1 mm. wide. Four anthers stand at the orifice and one within the tube. The stigma, which matures later than the anthers, also stands in the orifice.

INSECT VISITORS. In the Chiricahua Mt. population we observed the following flower visitors in August 1961. The butterflies and beeflies were getting nectar, and the bees were collecting pollen. *Ochlodes snowi* (Hesperiidae). *Papilio philenor* (Papilionidae). *Perdita sp.* (Andrenidae). *Dianthidium sp.* (Megachilidae). *Anthidium maculosum* (Megachilidae). *Villa sinuosa jaenickiana* (Bombyliidae).

SECTION MICROGILIA

Our third and last section of the genus Ipomopsis contains about 12 species of herbaceous plants, mostly low-growing, with small whitish flowers clustered in heads. The perennial species mostly occur in the mountains, often above treeline, or on cold northern plains, while the annuals inhabit the interior deserts.

Ipomopsis congesta   *Fig.* 19

Ipomopsis congesta is a low perennial herb which occurs on alpine fell-fields in the Sierra Nevada and adjacent mountain ranges and on the arid interior plains

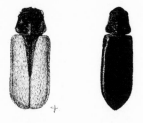

Fig. 19. Ipomopsis congesta, Trichochrous umbratus (Melyridae, upper left), and Eschatocrepis constrictus (Melyridae, upper right). Life size; the beetles are also shown enlarged.

east of the Sierra-Cascade axis. The small white flowers are grouped in terminal capitate heads 1.5 to 3 cm. broad. The flowering heads have a musky, spicy, or sweet odor in different races. The flowers are strongly protandrous and, on the basis of inconclusive tests, may be self-incompatible.

A variety of insects—beetles, small bees, wasps, butterflies, beeflies—visit the flowering heads for pollen or nectar. Many of these are casual as visitors, ineffective as pollinators, or both.

Trichochrous and other melyrid beetles are the most regular and numerous visitors and pollinators in six widely scattered populations observed. These small beetles with hairy bodies a few millimeters long crawl about over the flowering heads, eating pollen and nectar, and becoming covered with pollen grains. They are strong fliers and travel occasionally from plant to plant.

Small bees such as Chloralictus also bring about considerable pollination locally and occasionally, and so to a lesser extent do small beeflies like Pantarbes.

FLORAL MECHANISM. The small white flowers are congregated in terminal capitate heads. The size of the heads and the flowers and their type of odor vary geographically.

The race found on alpine fell-fields above treeline in the Sierra Nevada (*I. congesta montana*) consists of low cushion plants with large dense heads up to 3 cm. broad. The flowers give off a musky odor. The corolla is white with purple speckles and the anthers blue. The corolla tube is 5 or 6 mm. long in three Sierran populations studied, and the spreading limb is 4 or 5 mm. wide. The anthers are exserted about 1 mm. beyond the orifice and the stigma extends beyond the anthers. The flowers are strongly protandrous, the anthers drying up before the stigma is ripe.

The races inhabiting the interior plains in the Great Basin, Wyoming Basin, and Colorado Plateau (*I. congesta congesta*) consist of somewhat taller plants with smaller and frequently looser heads and sometimes with smaller flowers. These have a spicy odor in some populations and a faint sweet odor in others.

BREEDING SYSTEM. A strain of the alpine race from Sonora Pass in the Sierra Nevada has been tested in Claremont. The plants did not set seeds on selfing. Self-incompatibility is not indicated conclusively by these tests, however, because of the poor vigor of the plants in the Claremont environment and the difficulty of ruling out purely physiological causes of unfruitfulness.

INSECT VISITORS.

SONORA PASS AND SONORA PEAK, SIERRA NEVADA, CALIFORNIA (*I. c. montana*)

Trichochrous sp. (Melyridae). *Acanthoscelides lobatus* (Bruchidae). *Hylaeus sp.* (Colletidae). Unidentified wasp (Braconidae). Unidentified Hairstreak butterfly (Lycaenidae).

EBBETTS PASS, SIERRA NEVADA, CALIFORNIA   (*I. c. montana*)
   Trichochrous sp. (Melyridae).

BRIDGEPORT, MONO COUNTY, CALIFORNIA   (*I. c. congesta*)
   Chloralictus sp. (Halictidae). *Osmia sp.* (Megachilidae). *Ammophila sp.* (Sphecidae). *Melitaea acastus* (Nymphalidae). *Pantarbes sp.* (Bombyliidae).

SHIRLEY BASIN, CARBON COUNTY, WYOMING   (*I. c. congesta*)
   Unidentified melyrid beetles.

VERNAL, UINTAH COUNTY, UTAH   (*I. c. congesta*)
   Eschatocrepis constrictus (Melyridae). Another unidentified melyrid beetle.

Ipomopsis spicata   *Fig.* 20

   This low herbaceous plant of dry slopes and mesas in the eastern Rocky Mts. blooms in the spring, unlike its congeners, and bears its small yellowish-white flowers in cylindrical spikes. The flowers give off a cloying semen-like odor.

   The flowers are visited by nectar-feeding flies—tachinids and others—and by small pollen-collecting halictid bees, all of which cling and probe into the floral tube, and get pollen on various body parts in the process.

   The strong nitrogenous odor of the flowers of *I. spicata*, along with their small size, yellowish-white color, and aggregation in spikes, are unique in the Polemoniaceae, but are found in large segments of other families such as the Liliaceae, where the syndrome of floral characters is associated with fly pollination. The association of tachinid and other flies with *I. spicata* represents an interesting parallel development in one species of Polemoniaceae.

   FLORAL MECHANISM. Unlike other species of Ipomopsis, *I. spicata* blooms in the spring. The small yellowish-white flowers are borne in cylindrical spikes and give off a strong semen-like odor which is strongest in the morning. The salverform corolla has a nectar-containing tube 8 mm. long and about 2 mm. in diameter. The anthers and stigma are included midway in the tube.

   INSECT VISITORS. The following insects were seen in two populations in Colorado, one above Boulder and the other near Estes Park, in June 1962: *Chloralictus sp.* (Halictidae). *Augochlora sp.* (Halictidae). *Athanatus californicus* (Tachinidae). Another unidentified tachinid fly. Unidentified syrphid fly. *Bombylius lancifer* group (*Bombyliidae*).

The autogamous species   *Fig.* 21

   Eight species of the section Microgilia are annuals with small flowers which occur on the interior desert plains of North America (seven species) and South

Fig. 20. *Ipomopsis spicata and Athanatus californicus* (*Tachinidae*). *Life size.*
(*Fly is shown enlarged in Fig. 32 B.*)

Fig. 21. Ipomopsis pumila. Life size.

America (one species). Two of these are known from experimental tests and three others are inferred from examination of the floral mechanism to be autogamous.

Four strains of *Ipomopsis pumila* from Utah and Wyoming, and one strain of *I. polycladon* from the eastern Mojave Desert, set seeds autogamously in the screenhouse and produced fully vigorous I_1 progeny. Field observations of *I. depressa* in Nevada and of *I. minutiflora* along the Snake River in Oregon indicate that these plants are predominantly autogamous. *Ipomopsis gossypifera* in Chile and Argentina is similar to *I. pumila* and probably also autogamous.

7
ERIASTRUM AND
RELATED GENERA

Eriastrum, Langloisia, and Navarretia, considered in this chapter, along with Ipomopsis considered in the preceding chapter, comprise a group of related genera. The plants have some chromosomal and vegetative features in common, including a tendency to produce bristle-tipped or prickly leaves, and usually bloom in summer or late spring.

ERIASTRUM

Eriastrum is a genus of 14 species in western North America. One species, *E. densifolium*, is a subshrub, and the others are annual herbs. The flowers are funnelform, usually blue, commonly grouped in small wooly heads, and they bloom in summer. As compared with Ipomopsis and Gilia, Eriastrum is a fairly homogeneous genus in floral as well as vegetative features.

Eriastrum densifolium   *Plate I H*

The one perennial species of the genus, *E. densifolium*, occurs from the South Coast Ranges of California to Baja California and from the coast to the western Mojave Desert. It is a low prickly shrub with spreading stems terminating in showy blue flowering heads.

The funnelform flowers have a narrow throat in the interior mountain race and a broad throat in the coastal sand dune race. The stamens and style stand above the throat. The plants are predominantly outcrossing.

The flowers of the interior mountain race are visited by a variety of bees, beeflies, and butterflies, which perch on the heads and probe for nectar or collect pollen. The insect activity reaches its peak during the sunny part of the day and ceases after sundown. The most abundant and active flower feeders, and the main effective pollinators, are the native bees (Anthophora, Halictus, Lasio-glossum), and, in second place, the beeflies (*Bombylius lancifer*).

Numerous bumblebees (*Bombus vosnesenskii* group) have been observed visiting the broad-throated flowers of the coastal sand dune race. All bees were feeding on the nectar of Eriastrum (and getting pollen from Phacelia and Monardella in the vicinity). Their head fits into the full corolla throat and their 6 mm. long proboscis reaches down the tube, while pollen adheres to the venter and head.

FLORAL MECHANISM. The blue flowers are grouped in showy terminal heads. The corolla is funnelform—narrowly or broadly according to the race—with a limb 1.2 to 1.7 cm. wide, and a tube and throat 1.1 to 1.4 cm. long. The anthers and stigma stand several millimeters above the entrance to the nectar-containing tube. The flowers are strongly protandrous.

In the interior mountain race of southern California (*E. d. austromontanum*), as exemplified by populations in the San Bernardino Mts., the flowers are slender funnelform with a narrow throat 1 or 2 mm. wide at the orifice.

The coastal sand dune race of south-central California, by comparison, has broad funnelform flowers with an ample throat 4 mm. wide at the orifice.

BREEDING SYSTEM. A strain from Cajon Pass, San Bernardino County, was tested in Claremont. The plants were non-autogamous but did set seeds when artificially self-pollinated. The I_1 seeds did not germinate well, however, though the control seeds from sib crosses gave abundant germination. Outcrossing is promoted by the protandry and the apparently limited ability to produce seedlings by self-pollination.

INSECT VISITORS. The populations in the San Bernardino area listed below represent the interior mountain race of southern California. The Oceano population belongs to the coastal sand dune race of south-central California.

RUNNING SPRINGS, SAN BERNARDINO MTS.

Bees: *Halictus farinosus* (Halictidae). *Anthophora urbana* (Apidae). *Apis mellifera* (Apidae).

Beeflies (Bombyliidae): *Bombylius lancifer*. *Villa alternata*. *Villa fulviana*. *Villa morio*. Also an unidentified syrphid fly (Syrphidae).

Butterflies: *Vanessa cardui* (Nymphalidae). *Argynnis semiramis* (Nymphalidae). *Papilio rutulus* (Papilionidae). *Eurema sp.* (Pieridae). *Hesperia harpalus leussleri* (Hesperiidae).

Beetles: *Mordella albosuturalis* (Mordellidae).

Hummingbirds: *Calypte anna*.

SAN BERNARDINO MTS. (Merritt, 1897)

Solitary bees. Anthophora. *Apis mellifera*.

CAJON PASS, SAN BERNARDINO COUNTY

Bees: *Lasioglossum sisymbrii* (Halictidae). *Anthophora urbana* Api(dae). *Apis mellifera* (Apidae).

Beeflies (Bombyliidae): *Bombylius lancifer.*
Butterflies: *Pieris rapae* (Pieridae). *Lycaena gorgon* (Lycaenidae).

OCEANO, SAN LUIS OBISPO COUNTY
 Bombus vosnesenskii group (Apidae).

Eriastrum sapphirinum   Plate I J

This common annual species of dry open ground in southern California comes into bloom in early summer when the spring annuals have passed flowering. Its colonies form masses of blue color locally. The flowers are funnelform and medium small with a narrow throat and short tube about 6 mm. long.

The most numerous and consistent visitors and effective pollinators are small solitary bees (Perdita, Ashmeadiella). These bees may cling on the flowers to collect pollen or poke their head into the throat and their proboscis down the tube to suck nectar, getting pollen on the venter in either case.

Timberlake (1954) notes that *Perdita pelargoides* and two other species belonging to the same subgenus (Glossoperdita) have been collected only on flowers of Eriastrum and related genera of the Polemoniaceae. He states (p. 372): "The slender form of the bees and the long tongue are adaptations to obtain nectar and pollen from the long tubular flowers of the Phlox family." In the specimens of *Perdita pelargoides* collected on the Claremont population of *Eriastrum sapphirinum,* the proboscis is 3.5 mm. long.

FLORAL MECHANISM. The erect wiry stems bear the bright blue flowers at their tips. The flowers open in the warm sunny hours of morning and close up at dusk. The corolla is funnelform with exserted anthers and stigma. In the Claremont population the corolla tube is 6 mm. long and the orifice about 1 mm. wide. In other local races the tube may be shorter or the throat broader.

INSECT VISITORS. All localities are in southern California.

CLAREMONT, LOS ANGELES COUNTY
 Perdita pelargoides (Andrenidae). Perdita, two other species. *Diadasia sp.* (Apidae). *Apis mellifera* (Apidae).

SAN BERNARDINO MTS.
 Ashmeadiella sp. (Megachiliade). *Anthophora flexipes* (Apidae). *Apis mellifera* (Apidae). Hairstreak butterfly (Lycaenidae).

SAN BERNARDINO MTS. (Merritt, 1897, under "*Gilia virgata*")
 Anthophora sp. Apis mellifera. Butterflies.

SEVERAL LOCALITIES (Timberlake, 1954)
 Perdita pelargoides. Perdita tristella. (Andrenidae).

SEVERAL LOCALITIES   (Hurd and Michener, 1955, under "*Hugelia virgata*")
   Hoplitis colei. Hoplitis grinnellii. Ashmeadiella sonora. Ashmeadiella bigeloviae. Ashmeadiella gillettei cismontanica. Ashmeadiella californica californica. Ashmeadiella meliloti meliloti. (Megachilidae).

Eriastrum eremicum

Eriastrum eremicum is a low wiry annual of the Mojave Desert. It grows on sandy flats, frequently with *Gilia latiflora;* the latter blooms in April, but the Eriastrum flowers in May after the Gilia has gone to seed. The pale blue flowers are slightly zygomorphic with a corolla tube about 1 cm. long, which contains nectar in the morning, and exserted stamens and style.

In a population on the western Mojave Desert fair numbers of the beefly Pantarbes were observed to visit the flowers during the morning, while nectar was present, but not in the late afternoon when nectar was no longer in flow. The beeflies perch lightly on the flowers, insert their head in the orifice, and probe down the tube for nectar with their 4 mm. long proboscis. Their venter contacts the anthers and stigma and carries pollen.

FLORAL MECHANISM. The flowers are pale blue and slightly bilabiate with three erect upper corolla lobes and two spreading lower lobes. The corolla tube is about 1 cm. long. It contains nectar in midmorning but not usually in the hot dry afternoon. The stamens and style are exserted about 4 mm. beyond the orifice and on the lower side. The flowers open in the morning and close in late afternoon.

INSECT VISITORS. We observed two species of beeflies, chiefly *Pantarbes pasio* but occasionally *Lepidanthrax sp.*, visiting a population near Adelanto, San Bernardino County.

Eriastrum luteum   *Plate II*

This annual herb of the inner South Coast Ranges of California is unique in the genus for having bright yellow flowers. The small fragrant funnelform flowers are visited and pollinated chiefly by small solitary bees (Ashmeadiella, Dialictus), which fit well into the throat and carry much pollen on their venter or legs. Of definite but secondary importance as pollinators are small beeflies (Oligodranes, Phthiria).

FLORAL MECHANISM. The flowers are bright yellow and sweet-scented. The corolla is funnelform with a limb 1 cm. broad and throat 2 mm. wide at the orifice. The anthers and stigma are exserted beyond the throat.

INSECT VISITORS. The following insects were observed in a population near Creston, San Luis Obispo County, California: *Dialictus sp.* (Halictidae). *Ashmeadiella sp.* (Megachilidae). *Oligodranes sp.* (Bombyliidae). *Phthiria sp.* (Bombyliidae).

LANGLOISIA

The minor genus Langloisia, which is related to Eriastrum but is more advanced in various morphological characters, contains four species of small desert annuals with prickly leaves. The flowers are radially symmetrical in two of the species, but in the other two are distinctly bilabiate, an unusual condition in the family Polemoniaceae.

Langloisia punctata

This species of the Mojave Desert has large, erect, almost campanulate flowers with an open bell-shaped throat decorated with numerous purple dots. The nectar stored in the corolla tube is accessible through small holes in the base of the throat. The stamens and style stand up in a column in the center of the flaring throat. The flowers are protandrous and non-autogamous.

The flowers are visited and pollinated, among other more casual agents, by unidentified large beeflies (Bombyliidae), which hover to probe for nectar and in so doing contact the anthers and stigma with their face and chest.

FLORAL MECHANISM. The flowers are terminal on the stems, erect, open, almost campanulate, and radially symmetrical. The corolla is white or pale violet dotted with numerous small purple spots. There is a corolla tube 1 cm. long, mostly enclosed by the calyx, and containing nectar, and a broad open throat-like region formed by the ascending corolla lobes. This bell-shaped throat is about 1 cm. deep. Five small holes in its base, marked by purple nectar guides, lead down to the nectar chamber in the tube. A part of its wall is lined with a bright yellow succulent tissue composed of bulbous cells. The column of stamens and style stands in the center of the throat with the stigma raised above the anthers at maturity. The flowers are protandrous.

BREEDING SYSTEM. A strain from the southern end of Death Valley, California, was grown in breeding cages in Claremont. The plants are non-autogamous for the most part, though not completely so, in that one flower out of scores produced seeds within the cage. Observation of the floral mechanism indicates that the protandrous development and spatial separation of the anthers and stigma are normally sufficient to prevent automatic self-pollination, but are subject to exceptional lapses.

INSECT VISITORS. We have repeatedly observed two populations, one in Death Valley and the other in Cronise Valley, San Bernardino County, California. Insect activity on the flowers is slight, and despite much searching the pollination system of this species is still not well understood.

We have seen melyrid beetles sitting in the flowers, unidentified beeflies visiting them by day, and Celerio lineata (Sphingidae) making visitations by night. All of these insects could bring about some pollination.

The hawkmoth seems to be only a casual visitor, poorly adapted to feed on the Langloisia flowers. Whether the beetles are significant factors in the pollination of Langloisia or not cannot be decided as yet.

The most consistent visitor is the beefly, which has not been caught, partly because it is very fast and wary and partly because observing its flower-visiting behavior has had a higher priority than identifying it. It is a large, dark brown beefly with a wide wingspread and pictured wings, such as Poecilanthrax for example. It hovers and probes down the floral tube for nectar. This action brings its face or chest into contact with the anthers and stigma. The beeflies go rapidly from plant to plant.

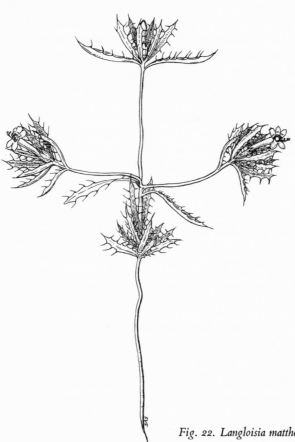

Fig. 22. Langloisia matthewsii and Pantarbes sp. (Bombyliidae). Life size. (Pantarbes is shown enlarged in Fig. 31 E.)

Langloisia matthewsii *Fig.* 22

This low tufted annual of the Mojave Desert and neighboring desert areas has horizontal bilabiate flowers with a tube 1 cm. long, a spreading limb divided into an upper and lower lip, and projecting stamens and style. The flowers are pink with reddish-purple speckles.

The most frequent and regular visitor in two populations observed is an ashy-white beefly (Pantarbes). This beefly probes for nectar in the tube with its 5 or 6 mm. long proboscis, getting pollen on its venter in the process, and flies actively from plant to plant. Bees (Anthophora) visit the flowers in fewer numbers and bring about some pollination too.

FLORAL MECHANISM. The flowers are horizontally disposed and bilabiate. The corolla tube is 1 cm. long and about 1 mm. wide. The corolla limb consists of three upper lobes and two lower ones. The limb is dull pink with a white center around the orifice of the tube and a zone of reddish-purple speckles on the upper lobes. The stamens and style project 7 or 8 mm. in front of the corolla limb.

INSECT VISITORS. Two populations in the El Paso Mts., Kern County, California, were observed. The most frequent visitor in both populations was *Pantarbes sp.* (Bombyliidae). *Anthophora sp.* was present in fewer numbers in one population.

NAVARRETIA

This genus of summer-blooming annuals has its main center of distribution in California, where 29 of the 30 species are found. Several of these species extend more widely throughout western North America, and one other occurs in southern South America. The flowers are small and funnelform and frequently blue or purple but sometimes pink, yellow, or white.

The genus is divided into four sections. The first three of these—Aegochloa, Masonia, and Mitracarpium—occur in open fields and meadows and have brightly colored flowers. In the section Aegochloa the plants are coarse, spiny-leaved, glandular, and skunky-scented; while in Masonia and Mitracarpium they are usually smaller with softer and more delicate foliage. The species for which pollination records are available belong to these three sections as follows: Aegochloa (*N. atractyloides, hamata, squarrosa*); Masonia (*N. peninsularis, viscidula*); Mitracarpium (*N. pubescens*).

The fourth section, Navarretia, contains eight species of specialized inhabitants of wet swales and vernal pools. The flowers are pale in color and range in size from small to minute with the anthers exserted or included. The modes of pollination in this section are unknown. However, it is probable that the species like *N. involucrata* and *N. subuligera* possessing minute flowers with included sex organs are autogamous.

Navarretia atractyloides Fig. 23

This prickly annual of the California foothills has small, narrowly funnelform, deep violet, nectariferous flowers. In a population in the blue-oak savanna of the La Panza Range, South Coast Ranges, we have observed visitations by two species of small bees, two species of medium-sized bees, and a small beefly. The small bees, Ashmeadiella and Oreopasites, which fit into the flowers nicely and go actively from plant to plant with Navarretia pollen on their venter, were the most numerous, consistent, and effective pollinators in this population. The beefly Lepidanthrax was also carrying out considerable pollination. In a population in the San Gabriel Mts. farther south, Timberlake (1954) has recorded flower visits by the small bee, *Perdita pelargoides*.

FLORAL MECHANISM. The flowers are small, funnelform, nectariferous, and deep violet with a yellowish center. In a population near La Panza mentioned below the corolla tube and throat are 1.7 cm. long, the orifice 1.5 mm. wide, and the corolla limb 5 mm. broad. The stamens stand at several levels from within the throat to just above the orifice, and the style is exserted at maturity.

INSECT VISITORS.
LA PANZA RANGE, SAN LUIS OBISPO COUNTY
Ancylandrena sp. (Andrenidae). *Ashmeadiella sp.* (Megachilidae). *Osmia sp.* (Megachilidae). *Oreopasites sp.* (Apidae). *Lepidanthrax sp.* (Bombyliidae).
LYTLE CREEK WASH, SAN BERNARDINO COUNTY (Timberlake, 1954)
Perdita pelargoides (Andrenidae).

Navarretia hamata

This species of low prickly annuals occurs in coastal southern California. The small violet funnelform flowers are visited by various small bees, beeflies, syrphid flies, and butterflies, The syrphid flies and butterflies extract nectar but carry little or no pollen. A small bee, Exomalopsis, effects some pollination. The most numerous and active in the one population studied were nectar-feeding beeflies (Bombylius, Lordotus, Lepidanthrax), and these insects were bringing about most of the pollination.

Fig. 23. *Navarretia atractyloides.*
Life size; flower enlarged.
(From Abrams, 1951.)

FLORAL MECHANISM. In a population in the Claremont area for which pollination records are available the flowers are funnelform and violet with a white-ringed purple eye. The corolla tube and throat are 7 or 8 mm. long, the throat 2 mm. wide at the orifice, and the limb 8 mm. wide. The stamens and style stand above the throat, the stigma exserted beyond the anthers.

INSECT VISITORS. The following visitors were observed in a population in the foothills of the San Gabriel Mts. near Claremont, Los Angele County: *Exomalopsis sp.* (Apidae). *Bombylius lancifer* (Bombyliidae). *Lordotus planus* (Bombyliidae). *Lepidanthrax sp.* (Bombyliidae). Unidentified syrphid fly. *Euphydryas chalcedona* (Nymphalidae).

Navarretia squarrosa

Navarretia squarrosa is a tall viscid herb with a skunk-like odor which occurs in the coastal region from north-central California to Washington. The small blue-violet flowers have included stamens and style, and seem to be self-pollinating to a large extent. Anthophora has been observed visiting the flowers at rare intervals, and other bees and perhaps beeflies may be expected to be found on the flowers with further observations. Such occasional visits lead to some cross-pollination.

FLORAL MECHANISM. The blue-violet funnelform corollas have a small throat about 1 mm. wide at the orifice and a tube and throat about 8 mm. long. The anthers and stigma are included in the throat, so that self-pollination is possible.

INSECT VISITORS. The flowers of a population on Pt. Reyes Peninsula, Marin County, were mostly not being visited by any insects during two days of observation. Individuals of Anthophora working mainly on neighboring plants of *Lupinus arboreus* and *Silene gallica* made occasional exploratory visits to the Navarretia but soon returned to their original food plants. The Navarretia plants seemed to be self-pollinating to a large extent during the observation period.

Navarretia peninsularis

These low plants with spreading branches and small funnelform flowers occur in the mountains of southern California. As the stigma and anthers ripen on the same level, the plants may be self-pollinating to some extent.

Three species of beeflies (Bombylius, Anastoechus, Lepidanthrax) have been observed feeding on the nectar of a population in the San Bernardino Mts. The Bombylius and Anastoechus are larger than the flowers and feed in a hovering position, carrying pollen on the face. The smaller Lepidanthrax corresponds in size to the Navarretia flowers and its head fits into the corolla throat. Numerous individuals of Lepidanthrax were actively feeding on the flowers and bringing about most of the pollination in the Navarretia population.

FLORAL MECHANISM. The flowers are pale whitish-violet with purple lines in the throat. The funnelform corolla has a tube 4.5 mm. long, a throat 2 mm. long and 2 mm. wide at the orifice, and a limb 3 or 4 mm. broad. The anthers are disposed on three levels from within to just above the throat. The stigma at maturity is in the zone of anthers. The flowers are protandrous, but not completely so, and self-pollination is possible.

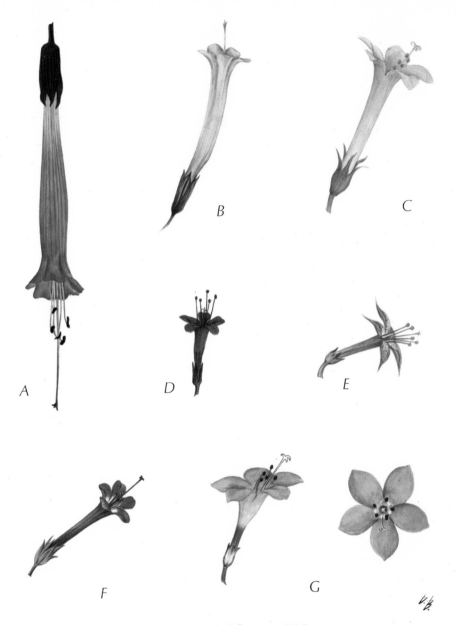

Plate II. Hummingbird flowers. All life size.

(A) Cantua candelilla.   (B) Huthia longiflora.   (C) Polemonium pauciflorum.
(D) Ipomopsis tenuifolia.   (E) Ipomopsis aggregata.
(F) Loeselia mexicana (as seen from above).   (G) Gilia splendens (San Bernardino race).

INSECT VISITORS. The junior author observed three kinds of beeflies visiting the flowers of a population in Holcomb Valley, San Bernardino Mts., in June 1962: *Bombylius lancifer. Anastoechus barbatus. Lepidanthrax sp.* The proboscis is 5 mm. long in the specimens of Bombylius and Anastoechus, and 2 mm. long in the Lepidanthrax.

Navarretia viscidula

Navarretia viscidula in the Sierra Nevada and North Coast Ranges of California has bicolored flowers with a blue or purple limb and a pale yellow throat and tube. The corolla tube ranges up to 9 mm. long. The anthers and stigma stand above the throat. Hurd and Michener (1955) found the megachilid bees, *Ashmeadiella californica californica* and *A. bucconis denticulata*, visiting the flowers near Mariposa in the Sierra Nevada.

Navarretia pubescens

Navarretia pubescens occurs in the valleys and foothills of central and northern California. The flowers are deep blue-violet with a tube 6 or 7 mm. long and throat 4 mm. long and 3 mm. wide. The stamens and style are well exserted. Small bees (*Andrena sp.*) have been observed visiting the flowers of a population in the Sierra Nevada foothills, Tulare County, carrying pollen on the venter from plant to plant.

8

LEPTODACTYLON AND LINANTHUS

Leptodactylon is a small genus of shrubs and Linanthus is a larger and hetero-geneous genus of herbaceous plants, mostly annuals. The six species of Leptodac-tylon and all but one of the approximately 38 species of Linanthus occur in western North America with a center of distribution in California. One species of Linanthus is found in Chile.

Leptodactylon and Linanthus share a number of common characteristics in the chromosome complement, leaves, and flowers. They form a third circle of affinity within the Gilia tribe, one which is advanced compared with the genus Gilia. The plants have palmate and usually opposite leaves and bloom in the spring. The showy flowers are brightly colored or white, are mostly sessile in small terminal clusters on the stems, and are open by day in most species.

The six sections of Linanthus fall into two phylogenetic branches: (1) Siphon-ella, Pacificus, Leptosiphon, Dactylophyllum; and (2) Dianthoides, Linanthus. The sections differ in floral characters related to pollination, as we shall see.

LEPTODACTYLON

Leptodactylon californicum   *Plate III B, Fig. 24*

This evergreen prickly shrub is common in the chaparral zone of the southern California mountains and occurs also on the sandy coastal plain of south-central California. It is floriferous for an extended period in the spring, making a bright pink spectacle which can be spotted easily by a human observer from a consider-able distance. The flowers are also very fragrant and scent the air all around the bush on a calm sunny day.

The salverform corolla consists of a broad, bright pink limb, 3 cm. or more in diameter, and a long slender tube, 1.2 to 1.5 cm. long, containing nectar. The

Fig. 24. Leptodactylon californicum, Papilio rutulus (Papilionidae), and Hemaris senta (Sphingidae). Life size.

anthers and stigma are included well within the tube. The plants of one strain tested are self-incompatible.

Three populations in the San Gabriel Mts. of southern California have been observed repeatedly in different years. A fourth population far away in the South Coast Ranges has also been watched. There is considerable regularity in the spectrum of visitors from year to year and from area to area.

Hummingbirds feeding in the area have never been observed to visit the Leptodactylon flowers, which are too slender to be probed easily by the bird bills. Solitary bees likewise pay little or no attention to the Leptodactylon flowers, or make occasional exploratory flights but do not pause long, as the nectar and usually also the pollen are buried in the tube below their reach. *Bombylius lancifer* visits the flowers with some regularity. Its tongue is long enough to reach the upper capillary nectar and to get pollen on the basal part in the process, but since the pollen-carrying regions of the tongue do not contact the stigma the beefly does not normally effect pollination.

The most consistent visitors and the effective pollinators of *Leptodactylon californicum* are butterflies, day-flying hawkmoths, and cyrtid flies. *Papilio rutulus*, *Hemaris senta*, and *Eulonchus smaragdinus* in particular have been found on the flowers in different years and in different areas. The butterflies and cyrtid flies settle down on the flowers to feed, while the hawkmoths hover, and in either case the long proboscis probes down the corolla tube for nectar. Leptodactylon pollen was found adhering to the distal parts of the proboscis of all specimens examined and was strewn down the floral tubes after their feeding visits.

The proboscis lengths of the three common visitors are: *Papilio rutulus*, ± 2.3 cm.; *Hemaris senta*, ca. 1.5 cm.; *Eulonchus smaragdinus*, 1.1 to 1.4 cm. The tongues of these nectar-feeding visitors thus correspond well with the length of the floral tube of *Leptodactylon californicum*. *Celerio lineata* with a proboscis ± 4 cm. long greatly exceeds the flower tube in length, and the body of this moth remains about 2 cm. above the flowers as it hovers to feed, but its tongue tip contacts the internal anthers and stigma and effects pollination.

FLORAL MECHANISM. The flowers are salverform, very fragrant, and bright pink, often with a white eye marking the orifice. The long slender corolla tube is 1.2 to 1.5 cm. long and 1 mm. wide. The broad corolla limb measures 3 cm. in diameter in the chaparral races and up to 3.5 cm. in the coastal sand dune race. Nectar collects at the base of the corolla tube. The anthers are included in the tube 2 or 3 mm. below the orifice and about 1 cm. above the base. The stigma, also included, rises to a level several millimeters below the anthers. The anthers and stigma are mature concurrently.

BREEDING SYSTEM. A strain from Santa Ana Canyon, Orange County, was artificially self-pollinated and sib-crossed, and proves to be self-incompatible.

INSECT VISITORS.

SAN GABRIEL MTS., LOS ANGELES COUNTY   (3 populations)

Papilio rutulus (Papilionidae). Unidentified Hairstreak (Lycaenidae). *Ochlodes sylvanoides* (Hesperiidae). *Hemaris senta* (Sphingidae). *Bombylius lancifer* (Bombyliidae). *Eulonchus smaragdinus* (Cyrtidae).

LA PANZA RANGE, SAN LUIS OBISPO COUNTY

Papilio rutulus (Papilionidae). Unidentified Dusky-winged skipper (Erynnis or Thorybes). *Hemaris senta* (Sphingidae). *Proserpinus clarkiae* (Sphingidae). *Celerio lineata* (Sphingidae).

LINANTHUS SECTION SIPHONELLA

The section Siphonella of Linanthus contains two species of perennial herbs and one annual with funnelform flowers in western North America.

Linanthus nuttallii

This bushy perennial herb has a widespread distribution in the mountains of western North America. The fragrant white flowers are borne in terminal clusters around the periphery of the plant. They are salverform with a tube 6 or 7 mm. to 1 cm. long. The anthers stand in the narrow orifice and the stigma in the tube just below.

During the daytime the flowers receive slight and sporadic visitations by skippers, beeflies, and other insects, which are relatively ineffectual as pollinators. By night various small moths (geometrids, pyralids) are active around the plants.

FLORAL MECHANISM. The flowers are white and give off a sweet fragrance, chiefly by night in some races. The corolla is salverform with a broad spreading limb, a tube ranging from 0.6 to 1.0 cm. long in different races, and a narrow orifice. The anthers stand at or near the orifice and the stigma is usually in the tube below the anthers.

INSECT VISITORS. The following localities are in California.

SAN BERNARDINO MTS. (by day)

Erynnis propertius (Hesperiidae). *Erynnis callidus* (Hesperiidae). *Bombylius lancifer* (Bombyliidae).

Eclimus luctifer (Bombyliidae).   Unidentified syrphid fly.   *Chaetogaedia monticola* (Tachinidae). *Anthaxia aeneogaster* (Buprestidae).   Unidentified melyrid beetle.

BIG PINE CREEK, SIERRA NEVADA, INYO COUNTY   (by night)

   Neoterpes trianguliferata costinotata (Geometridae).   *Ephestiodes gilvescentella* (Pyralidae).   *Crambus pascuellus* (Pyralidae).

SECTION PACIFICUS

   This section, containing the single species *L. grandiflorus*, is closely related to the preceding section, Siphonella, and like it is characterized by funnelform flowers.

Linanthus grandiflorus   *Fig. 25*

   This annual species of the central California coast has fragrant white funnelform flowers borne in small clusters. The flowers are open by day but close at night. The throat is large and slopes down to a short tube about 5 mm. long. The orange anthers and stigma stand at the orifice. The plants are self-compatible but usually non-autogamous.

   The main pollinators are bumblebees. Fairly large bees belonging to the *Bombus vosnesenskii* group have been seen visiting the flowers actively and in fair numbers on Pt. Reyes Peninsula. The bees poke their heads into the corolla throat, which accommodates them well, and collect pollen or nectar as the case may be, getting pollen on various body parts which come into contact with the stigma.

   FLORAL MECHANISM. The flowers are white or pale lavender, fragrant, large, and are borne in small terminal clusters. They are open by day and close at night. The corolla lobes and throat form a wide cup, sloping down below to the tube, which is about 5 mm. long. The orange anthers stand at the orifice and the stigma below the anthers. The flowers are incompletely protandrous.

   BREEDING SYSTEM. A strain from Pt. Reyes, Marin County, has been observed and tested in Claremont. The plants are self-compatible but are predominantly non-autogamous. The I_1 seedlings were as vigorous as the control seedlings from sib crosses.

   INSECT VISITORS. We observed the following insects visiting the flowers of the Pt. Reyes population: *Bombus vosnesenskii* group (Apidae). *Vanessa sp.* (Nymphalidae). *Serica anthracina* (Scarabaeidae). The beetle was chewing the corolla limb and the butterfly extracting nectar without effecting any significant amount of pollination.

   Fair numbers of bumblebees were entering the broad corolla throat to collect pollen or nectar. The pollen collectors had pollen on the forelegs. The bees taking nectar insert a tongue about

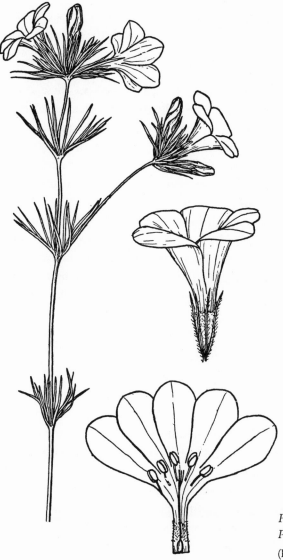

Fig. 25. Linanthus grandiflorus.
Plant life size; flowers enlarged.
(From Abrams, 1951.)

7 mm. long down the corolla tube, getting pollen around the mouth and on the venter. Some of these body parts normally come into contact with the stigma. As the bumblebees were actively and systematically working the Linanthus population, they were bringing about the great bulk of the pollination.

SECTION LEPTOSIPHON

The section Leptosiphon of Linanthus contains about 10 species of annuals distributed on the Pacific slope of North America with the center of distribution in California. The plants form colonies at the edges of meadows or in openings in pine woods or in the open foothills.

The general color effect of the bright-hued flowers in these colonies is a massive pink, lilac, white, or yellow. The individual flowers are usually bicolored or tricolored, often with a yellow center contrasting with the pink, lilac, or white of the limb, and this condition breaks the evenness of the color display. In addition, the populations are frequently polymorphic for flower color, containing white as well as colored plants, the mixture of which further increases the brightness of the overall color effect of the colony.

The flowers are clustered in small heads at the ends of the branches and are salverform. Their most remarkable and distinctive feature is the long slender corolla tube, which is 1.5 to 2.5 cm. long in most of the species and reaches the extreme length of 4 cm. in some. The nectar in the lower part of this tube is normally accessible only to long-tongued flies, which are the chief agents of pollination.

Linanthus parviflorus   Plate III E

Linanthus parviflorus grows in grassy or wooded foothills throughout the length of California. The salverform flowers have a small white limb and a long filiform tube. This tube varies from 1.5 to 3.0 cm. long in the populations for which pollination records are available. Nectar collects and rises in the lower part of the tube, and the anthers and stigma stand just above the orifice. The plants are self-compatible but non-autogamous.

In three populations the most frequent and consistent visitor was *Bombylius lancifer*. This beefly settles down lightly on the corolla limb, maintaining a partially hovering position with its wingbeat, and inserts its slender straight proboscis down the corolla tube for nectar. The proboscis ranged from 8 to 10 mm. long in the specimens taken on *Linanthus parviflorus*, which brings the upper capillary nectar within its reach. The beeflies had yellow Linanthus pollen adhering to the face and sometimes to the legs. They fly actively and systematically through the plant population.

The cyrtid fly, Eulonchus, with a proboscis longer than that of Bombylius, has not yet been seen on *Linanthus parviflorus*, but could be expected on the local races with very long tubes in view of its known association with other long-tubed species in the section.

FLORAL MECHANISM. The salverform flowers have a narrow white limb about 1 cm. in diameter, usually with a yellow center, and a long filiform tube less than 1 mm. wide. The length of the tube varies in different local races. In the Sequoia Park population listed below it is 1.5 to 2.0 cm. long; 2.5 cm. in the Lake Berreyessa population; 2.5 to 3.0 cm. in the Angwin population; and even longer (to 4 cm.) in some other populations for which pollination records are not yet available.

Nectar collects in the base of the tube by day and may rise as high as 1 cm. below the orifice. The anthers stand just above the orifice and the stigma above the anthers. The flowers are open on sunny days but close up at night by the folding of the lobes and remain closed during cold days.

BREEDING SYSTEM. A strain from Lake Berreyessa, Napa County, is self-compatible but non-autogamous. The I_1 progeny were as vigorous in their general growth as the progeny of sib crosses, but exhibited a significantly higher proportion of floral abnormalities in hot dry weather.

Darwin (1877) considered the possibility of heterostyly (in "*Gilia micrantha*") on the basis of a few specimens received from Kew with styles varying from long to short. There is considerable variation in flower length, breadth, and coloration in many populations of *Linanthus parviflorus*, but it does not seem to represent true heterostyly.

INSECT VISITORS.

LAKE BERREYESSA, NAPA COUNTY
   Bombylius lancifer (Bombyliidae).

ANGWIN, NAPA COUNTY
   Bombylius lancifer (Bombyliidae). *Andrena sp.* (Andrenidae). *Adela simpliciella* (Adelidae). Unidentified melyrid beetle.

SEQUOIA NATIONAL PARK, TULARE COUNTY (David P. Gregory)
   Bombylius lancifer (Bombyliidae). *Villa agrippina* (Bombyliidae).

Linanthus androsaceus *Fig. 26*

Linanthus androsaceus occurs in the Coast Ranges and along the coastline of central and northern California. In the Coast Ranges it frequently grows in the same hills as *L. parviflorus* but in a different altitudinal zone. The two species hybridize locally, which contributes to the variability observed in some populations. The coastal mesas are inhabited by the form known as *L. androsaceus croceus* or *L. rosaceus*, which is probably but not certainly a race conspecific with the Coast Range populations.

Fig. 26. *Linanthus androsaceus croceus and Eulonchus smaragdinus (Cyrtidae).* Life size.

    The salverform flowers have a broad limb, which is frequently bright pink or violet, and a long filiform tube 2 to 3 cm. long. The plants are self-compatible and partially autogamous.

    The cyrtid fly, *Eulonchus smaragdinus*, with a needle-like bill 2 cm. long, has been observed visiting the flowers for nectar, and carrying pollen on its venter.

    FLORAL MECHANISM. The flowers are borne in terminal heads and are open on warm sunny days but closed at night and during cold weather. They are salverform, as in *L. parviflorus*, but the limb is broader and is often bright pink or violet rather than white. The filiform corolla tube is 2 to 3 cm. long in the species and varies from 2.5 to 3.0 cm. in the population on Pt. Reyes mentioned below. The anthers and stigma stand just above the orifice. The flowers are incompletely protandrous with overlapping male and female stages.

    BREEDING SYSTEM. A strain from Pt. Reyes Peninsula, Marin County, is self-compatible. It is partially autogamous, in that some untouched flowers set capsules in the breeding cage but others did not. Darwin also found that caged plants set numerous capsules in the absence of insect visits (1878, ch. 10). The I_1 progeny of the Pt. Reyes strain were comparable in vigor to the progeny of sib crosses.

INSECT VISITORS. We observed *Eulonchus smaragdinus* (Cyrtidae) visiting the flowers of the population on Pt. Reyes. The fly had a proboscis 2 cm. long, which enabled it to reach the nectar in the lower part of the very long floral tube. It also had Linanthus pollen scattered thinly over the entire ventral surface of its thorax and abdomen.

Linanthus bicolor

This widespread species of the Pacific slope has salverform flowers with a long filiform tube and a narrow pink limb with a yellow center. A population growing sympatrically with *Gilia tricolor* has been observed in the blue-oak savanna of the foothills in Colusa County, California. Here the filiform corolla tube is 2 cm. long. A beefly, *Bombylius sp.*, was feeding on both the Linanthus and the Gilia and was carrying both types of pollen on its head.

Linanthus breviculus

This species has a localized distribution in the mountains of southern California where it grows in openings in dry pine woods and on the adjacent desert slopes. Its slender salverform flowers are slightly fragrant and usually white with a yellow eye. The filiform tube varies from 1.8 to 2.5 cm. long in a population in the San Bernardino Mts. The most abundant visitor observed in this population, and the chief pollinator of its flowers, is *Bombylius lancifer*.

FLORAL MECHANISM. The salverform flowers are fragrant and nectariferous by day and have a white limb with yellow center in most individuals. A population in Holcomb Valley in the San Bernardino Mts. consists chiefly of white-flowered plants but contains also a few pink and lavender-flowered individuals. The corolla tube in this population varies from 1.8 to 2.5 cm. long. The anthers stand at the orifice and the stigma 3 or 4 mm. above the anthers. This spatial separation and the protandrous development of the sex organs makes automatic self-pollination unlikely.

BREEDING SYSTEM. The floral mechanism tends to promote outcrossing to some extent as noted above. Merritt (1897) has suggested heterostyly in a white-flowered "*Gilia micrantha*" in the San Bernardino Mts., which is probably the species now known as *Linanthus breviculus*. She noted in a large sample of individuals that some were long-styled and others short-styled. The populations of *Linanthus breviculus* and of other species in the same group are indeed polymorphic for flower length and the correlated style length. This variation is not fully understood, but does not appear to fall into clear-cut alternative classes as in true heterostyly.

INSECT VISITORS. Two kinds of beeflies were visiting the flowers in the Holcomb Valley population. The most numerous visitor was *Bombylius lancifer* with a proboscis 8 to 10 mm. long in different individuals collected and with much Linanthus pollen on its face and around the base of the proboscis. It was carrying out most of the pollinatioL. A few individuals of *Villa alternata* with a short proboscis and some pollen on the venter were also visiting the flowers.

SECTION DACTYLOPHYLLUM

This group of 12 species occurs widely throughout western North America and in Chile. The flowers are distinctive in being small, cup-shaped, and solitary on long slender peduncles. They conform to the type of small-bee flower with filiform stalks which Vogel (1954, 45 ff.) has described in the South African flora.

Linanthus liniflorus   *Fig. 27*

The small-flowered form of this species, *L. liniflorus pharnaceoides*, which is the only one which concerns us here, has a wide distribution in the foothills of the Pacific slope. The plants are wiry annuals with small, pale blue, campanulate flowers borne on long slender stalks about 1 cm. long. The cup-like corolla throat, about 1 cm. wide above, slopes down to the short nectar-containing tube which is 3 or 4 mm. long.

The flowers are visited and pollinated chiefly by various small bees (Dufourea, Chelostoma, Ashmeadiella, etc.). The corolla throat accommodates their size and the slender peduncle supports their light weight. They can reach the nectar in the short corolla tube with their equally short proboscis or they gather pollen. In any case the orange Linanthus pollen adheres to their bodies and comes into contact with the stigma when they fly into another flower. Some supplementary pollination is carried out by small beeflies.

FLORAL MECHANISM. The following notes are based on three populations of the small-flowered race, *L. liniflorus pharnaceoides*, in south-central and southern California. The flowers are campanulate and pale blue or violet with a yellow center and fine purple veins converging toward the center. They are solitary on long slender peduncles 7 to 10 mm. long. They are open from early morning to midafternoon or late afternoon, at which time they close up.

The small cup-like throat is about 1 cm. in diameter at its widest part and slopes down to the short tube, which is 3 or 4 mm. long. The nectar in the tube is covered from above by a ring of hairs at the base of the throat. The stamens and style stand above the open throat with the stigma elevated above the anthers.

INSECT VISITORS. We have observed various small bees and other small insects on the flowers in three populations as noted below. Large bees cannot perch on the flowers because of the inadequate support afforded by the long slender peduncles, and usually do not try.

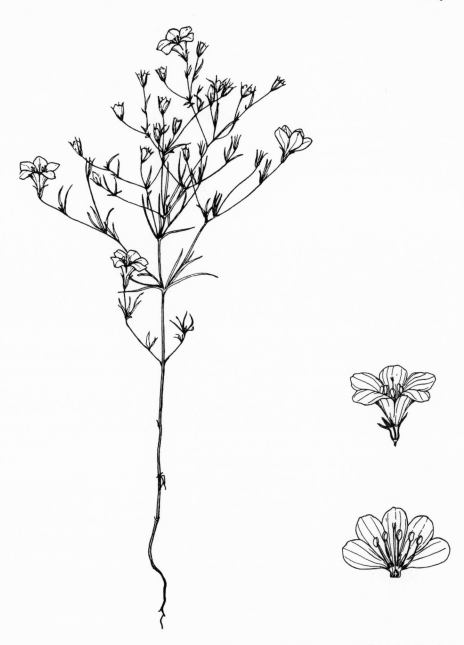

Fig. 27. *Linanthus liniflorus pharnaceoides, Ashmeadiella sp. (Megachilidae, upper left), and Dufourea linanthi (Halictidae, upper right). Life size; flowers enlarged. (Bees are shown enlarged in Fig. 30 E, G.)*

(Rearranged from Abrams, 1951.)

SAN GABRIEL MTS., LOS ANGELES COUNTY

Chelostoma sp. (Megachilidae). *Ashmeadiella sp.* (Megachilidae). *Calanthidium sp.* (Megachilidae). Halictus subgenus Seladonia (Halictidae). Ceratina, two species (Apidae). Unidentified small wasp (Braconidae). *Phthiria sp.* (Bombyliidae). Unidentified small fly (Conopidae). Two species of Melyridae.

LA PANZA RANGE, SAN LUIS OBISPO COUNTY

Dufourea linanthi (Halictidae). *Dufourea sp.* (Halictidae). *Oreopasites sp.* (Apidae). *Anastoechus sp.* (Bombyliidae).

CRESTON, SAN LUIS OBISPO COUNTY

Dufourea linanthi (Halictidae).

The autogamous species

Nearly half of the species of section Dactylophyllum have minute flowers 2 to 6 mm. long and may be supposed to reproduce autogamously. A strain of *Linanthus pygmaeus* from Guadalupe Island off Mexico, grown from seed collected by Dr. Sherwin Carlquist, sets a full complement of seeds by automatic self-pollination. The related *L. pusillus* in Chile, *L. harknessii* in the high Sierra Nevada, and *L. septentrionalis* in the Great Basin and northern Rocky Mts. are probably likewise predominantly autogamous.

SECTION DIANTHOIDES

The plants belonging to this section are low tufted annuals of the southwestern deserts and bordering regions with diurnal campanulate flowers.

Linanthus dianthiflorus   Plate I L

This species is fairly common in the dry valleys of southern California. Its erect campanulate flowers are pink with a yellow center marked by reddish spots. The flowers are visited and pollinated chiefly by small solitary bees (Perdita, Nomadopsis, Dufourea). Some supplementary pollination is effected by melyrid beetles.

FLORAL MECHANISM. The erect campanulate corolla is about 2 cm. wide across the top, and narrows down below to the short tube. The outer margin of the corolla lobes is fringed. The lobes are a soft pinkish-violet decorated with dark reddish spots at their bases and further set off by a yellow center in the throat. The stamens are exserted beyond the throat and the stigma is elevated slightly above the stamens.

INSECT VISITORS. The following insects have been observed in a population near Claremont, Los Angeles County: *Perdita sp.* (Andrenidae). *Nomadopsis barbata* (Andrenidae).

Dufourea sp. (Halictidae). *Trichochrous sp.* (Melyridae). The melyrid beetles are commonly found sitting in the flowers and bring about some pollination in their occasional flights from plant to plant. The small bees enter the flowers, getting pollen all over their venter and legs, and as they fly from plant to plant more frequently and actively than the beetles and carry more pollen, they are more effective pollinators of the *Linanthus dianthiflorus* flowers.

Linanthus parryae   *Fig. 28*

This species occurs on the Mojave Desert and in the inner South Coast Ranges. The campanulate flowers are blue or white in the western races and ivory-yellow in the eastern desert race. Populations on the western Mojave Desert frequently contain blue-flowered and white-flowered individuals intermixed in proportions that have remained constant during 20 years of observation and sampling (Epling, Lewis, and Ball, 1960). There are also some local all-blue populations. The plants are self-incompatible (Epling, Lewis, and Ball, 1960).

The campanulate flowers are sessile and erect on the low tufted stems and are open by day but closed at night. The open throat slopes down to a whorl of purple papillate ridges. Immediately below this whorl is a false tube, and just above it are the yellow anthers and stigma.

The flowers are visited by the small melyrid beetle, Trichochrous, which lands on the limb, walks around in the throat or takes shelter there, and sometimes pokes into the false tube or the zone of papillate tissue. In its movements within the flower it gets the Linanthus pollen on its body, which it carries on its occasional flights to neighboring plants.

FLORAL MECHANISM. The flowers are erect, campanulate, about 2 cm. wide across the top, and open by day but closed at night. In the western races, which are the only ones studied so far from the standpoint of pollination, they are blue or white with a yellow eye. Many populations in the western Mojave Desert are polymorphic for flower color, containing blue and white individuals in constant frequencies, while some populations are monomorphic (Epling, Lewis, and Ball, 1960).

The corolla lobes and throat funnel down to a false tube which extends from the level of the anthers to the point of insertion of the filaments. This false tube is marked by ridges of the corolla wall on both its upper and lower ends. The uppermost whorl of ridges consists of fine papillate purple tissue, while the lower whorl of ridges covers the ovary and the corolla tube proper except for five very narrow slots.

The yellow anthers are clustered around the central style at the base of the funnel just above the entrance to the false tube. The stigma is elevated above the anthers but not completely separated from them.

Fig. 28. Linanthus parryae and Trichochrous sp. (Melyridae). Life size; beetle also shown enlarged.

BREEDING SYSTEM. A strain from the western Mojave Desert has been found to be self-incompatible (Epling, Lewis, and Ball, 1960).

INSECT VISITORS. Epling, who has studied the population dynamics of *Linanthus parryae* in the western Mojave Desert for many years, has found the regular visitor and pollinator to be the small melyrid beetle, *Trichochrous sp.* (Epling, Lewis, and Ball, 1960). We have also seen Trichochrous repeatedly in the flowers of two populations in the western Mojave Desert and in another population on the desert slopes of the southern Sierra Nevada.

SECTION LINANTHUS

The four species comprising this group occur in the southwestern deserts and bordering arid regions. They have white vespertine flowers which, in the large-flowered species at least, are such as to attract moths.

Linanthus dichotomus   *Fig.* 29

Linanthus dichotomus ranges widely across the Mojave and Sonoran deserts and into the arid interior valleys of cismontane California. The flowers when open, which is exclusively at night in most races, expose a broad white funnel-shaped limb and give off a sweet fragrance. The limb slopes down to the nectar-containing tube, which is about 1 cm. long. The anthers and stigma are included in the middle part of this tube. The plants are self-compatible and capable of some autogamous seed formation. The flowers are occasionally visited by the hawk-moth, *Celerio lineata*, which hovers and probes into the corolla tube for nectar, and carries pollen on its tongue tip.

FLORAL MECHANISM. The flowers are sessile and erect on the upper branches. When open they are white and fragrant. The flaring white corolla lobes form a funnel 2.5 to 3.5 cm. wide at the top, sloping down to the corolla tube, which is about 1 mm. wide at the orifice and 8 to 10 mm. long. The tube contains nectar at the base. The sex organs are included midway in the tube with the stigma below the anthers.

In the widespread race of the southern California valleys and the southwestern deserts (*L. d. dichotomus*), the flowers are closed by day but open by night. In their closed position, the corolla lobes are furled up, exposing a dirty white or purplish external surface. At dusk the lobes unfurl in a matter of minutes, revealing the shiny white inner surface, and giving off a strong fragrance. By sunrise of the following morning the flowers are closed and odorless again. In the race of the North Coast Ranges (*L. d. meridianus*) the flowers remain open and fragrant during the day.

BREEDING SYSTEM. Strains of the northern race from Middletown, Lake County, and of the southern race from Bodfish, Kern County, have been found to be self-compatible. They are partially autogamous, in that some but not all untouched flowers set seeds. The I_1 seedlings showed no depression in vigor as compared with the progeny of sib crosses.

INSECT VISITORS. The flowers seem to be visited only rarely by insects. We have repeatedly watched different populations of the southern race by night in favorable weather without seeing any insect visits. Eastwood (1893) had a similar experience. A population of the northern race near Middletown, Lake County, was observed by day without success. At dusk, however, *Celerio lineata* began to feed on the flowers. The moth, with Linanthus pollen adhering to the distal part of its proboscis, flew rapidly and repeatedly from plant to plant.

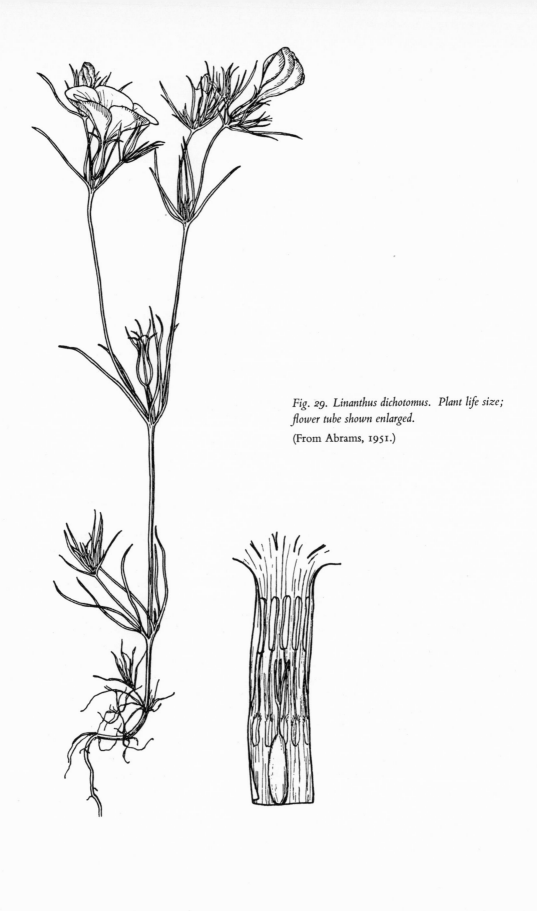

Fig. 29. Linanthus dichotomus. Plant life size;
flower tube shown enlarged.

(From Abrams, 1951.)

Fig. 30. Western American bees. Enlarged, not to same scale.

(A) Bombus edwardsi (Apidae).　　　(B) Tetralonia sp. (Apidae).

(C) Osmia distincta (Megachilidae).　　(D) Anthophora urbana (Apidae).

(E) Dufourea linanthi (Halictidae).　　(F) Chelostomopsis rubifloris (Megachilidae).

(G) Ashmeadiella sp. (Megachilidae).

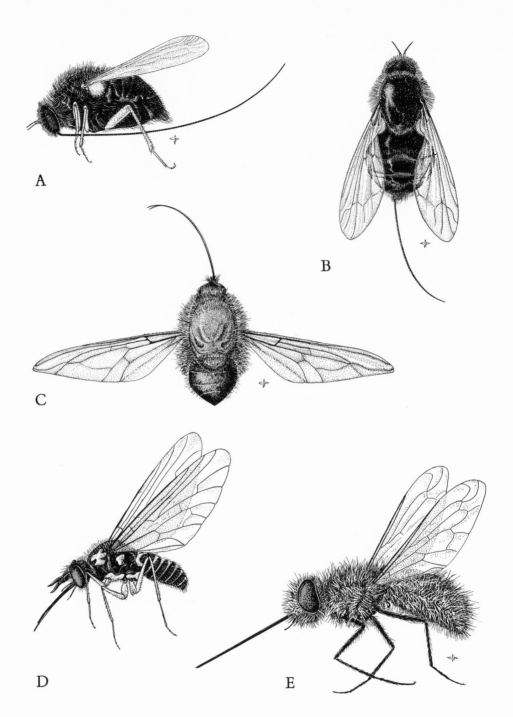

Fig. 31. Western American long-tongued flies. Enlarged, not to same scale.

(A, B) Eulonchus smaragdinus (Cyrtidae).   (C) Bombylius lancifer (Bombyliidae).
(D) Oligodranes sp. (Bombyliidae).   (E) Pantarbes pasio (Bombyliidae).

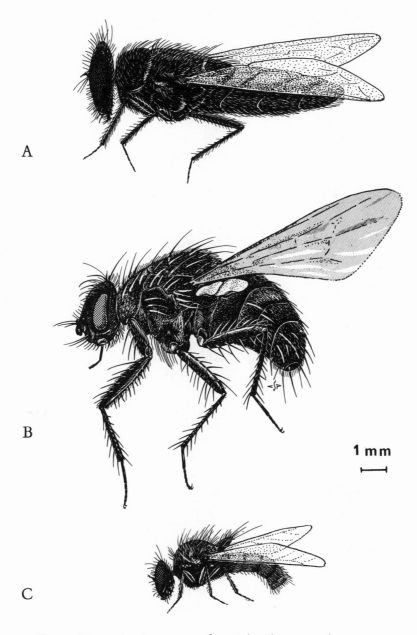

Fig. 32. Western American scavenger flies. Enlarged, to same scale.

    (A) Protophormia terraenovae (Calliphoridae).
    (B) Athanatus californicus (Tachinidae).
    (C) Thricops septentrionalis (Muscidae).

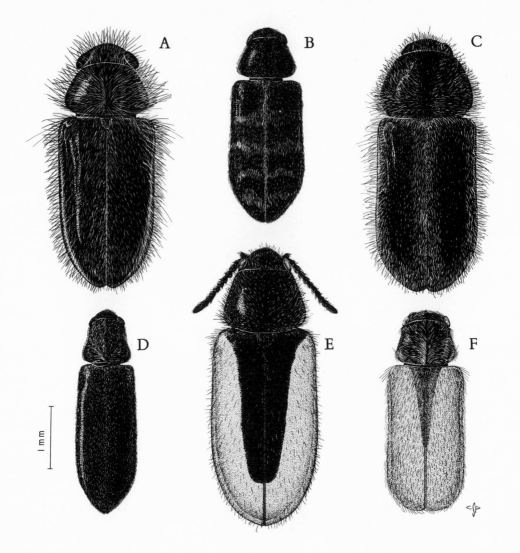

Fig. 33. Western American flower beetles (Melyridae). Enlarged, to same scale.

(A) Pristoscelis grandiceps. (B) Listrus famelicus. (C) Trichochrous suturalis.
(D) Eschatocrepis constrictus. (E) Listrus sp. (F) Trichochrous umbratus.

A

B

Fig. 34. Swallowtail butterflies (Papilionidae). Life size.
(A) Papilio rutulus.   (B) Papilio philenor.

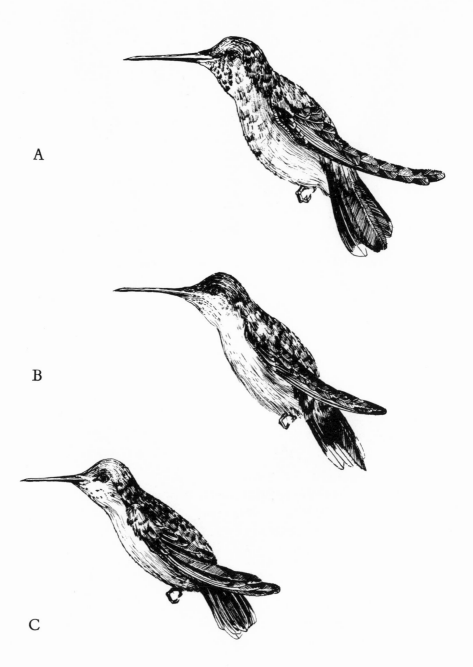

A

B

C

Fig. 35. Western American hummingbirds (Trochilidae). Life size.
(A) Calypte anna.   (B) Selasphorus rufus.   (C) Stellula calliope.

9

THE TROPICAL GENERA

Five genera of the Phlox family occur in the American tropics and subtropics from Mexico to Peru. These genera fall into three tribes which are rather distantly related to one another and are separated by wide phylogenetic gaps. The tribes and their constituent genera are: the Bonplandieae (Loeselia, Bonplandia); Cantueae (Cantua, Huthia); and Cobaeeae (Cobaea). Each tribal group presents a different ensemble of primitive and advanced features in the flowers as well as in the vegetative parts. Our knowledge of the reproductive biology of these interesting plants is fragmentary but suggestive.

LOESELIA

This genus of shrubby or herbaceous plants ranges through Mexico to northern South America. Most of the species, such as *L. glandulosa* and *L. coerulea*, have horizontal, bilabiate, blue or yellow flowers. The exserted arching stamens, style, and lower corolla lobes of these flowers afford a perching place for insects, probably bees. *Loeselia grandiflora* deviates from this condition in having white flowers with a slender tube and broad limb. Its flowers seem to be fitted for visits by hovering moths. *Loeselia mexicana* possesses red trumpet-shaped hummingbird flowers.

Loeselia mexicana   Plate II F, Fig. 38 C

This plant with a mound-like growth form begins to bloom after the end of the rainy season in Mexico and is in flower from October to February or March. Plants grown in southern California preserve the same periodicity. The numerous arching stems which arise from the woody base become clothed with bright green foliage throughout their length and with bright red flowers in the upper parts.

The red flowers are horizontally disposed and trumpet-shaped with a broad

Fig. 36. Long-billed Andean hummingbirds (*Trochilidae*). *One-half life size.*

(A) Patagona gigas peruviana.   (B) Coeligena torquata omissa.   (C) Ensifera ensifera.

Fig. 37. American flower-visiting bat (Phyllostomidae, Glossophaginae).
Leptonycteris nivalis. Life size.

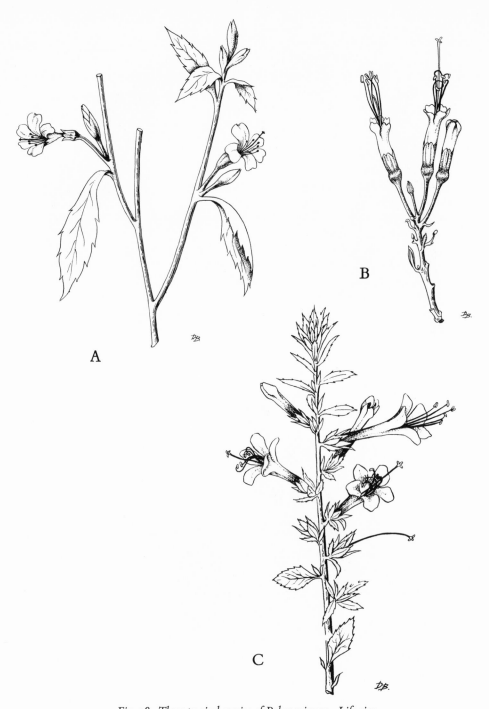

Fig. 38. Three tropical species of Polemoniaceae. Life size.

(A) Bonplandia geminiflora.   (B) Cantua pyrifolia.   (C) Loeselia mexicana.

tube about 2 cm. long and exserted stamens and style. The plants are strongly protandrous and self-incompatible. The flowers are visited by hummingbirds.

FLORAL MECHANISM. The following notes are based on plants grown in Claremont from seeds collected by Dr. J. Rzedowski in San Luis Potosí, Mexico. Herbarium specimens from other localities have similar floral characteristics.

The flowers are bright red, trumpet-shaped, and stand out horizontally on the plants. The floral tube is 2 cm. long and 2 mm. wide at the orifice. There is a constriction and slight bend in the tube near the base which separates the nectar chamber from the open part of the tube. The corolla lobes flare out to form a slightly bilabiate limb about 1.5 cm. in diameter which is red with a white or cream center.

The red stamens and style run along the lower wall of the corolla tube, extend well beyond the limb, and bend up at the tip so as to position the anthers or stigma on the lower side of the entrance to the floral tube. The flowers are strongly protandrous. As the stamens pass maturity the filaments retract, bringing the dried anthers closer to the orifice, while the style elongates simultaneously and the stigma opens in the position previously occupied by the anthers.

BREEDING SYSTEM. The San Luis Potosí strain has been tested in Claremont and found to be self-incompatible.

ANIMAL VISITORS. Mr. D. E. Breedlove (personal communication) observed a pair of hummingbirds feeding on the flowers of a population in Sinaloa in December 1962. It is also significant that the species has at least one Indian name and three Spanish names which mean hummingbird flower (huitzitzil-xochitl, flor de colibri, flor del chupamirto, chuparosa; Standley, 1924, 1211).

BONPLANDIA

Bonplandia geminiflora   Plate I A, Fig. 38 A

This leafy shrub occurs in moist tropical woodland in central and southern Mexico where it blooms from October to March. A strain from Sinaloa is self-compatible. The deep blue-violet flowers stand out horizontally and are bilabiate. The long corolla tube contains nectar, and the extended lower lip and exserted stamens and style afford a perching place for flower-visiting insects, perhaps for long-tongued tropical bees.

FLORAL MECHANISM. The following observations on plants grown in Claremont from seeds collected by Mr. D. E. Breedlove near Rosario, Sinaloa, may be taken as representative of the species.

The deep blue-violet flowers stand out laterally in pairs in the upper branches. They are subsalverform and bilabiate. The corolla tube is 2 cm. long and sharply bent below and is 3 mm. wide at the orifice. Nectar rises a short distance in the tube. The corolla limb consists of two reflexed upper lobes and three lower lobes which stand out as a lower lip. The stamens and style are exserted about 1 cm. beyond the orifice and lie on the lower lip. The flowers are incompletely protandrous.

BREEDING SYSTEM. Some preliminary tests made by Mr. Robert Rutherford with the Rosario strain indicate that the plants are self-compatible.

CANTUA

The genus Cantua contains four species of shrubs and small trees in the Andean region of Ecuador, Peru, and Bolivia. The trumpet-shaped flowers are white in *C. quercifolia* and *C. pyrifolia*, and reddish and very large in *C. candelilla* and *C. buxifolia*.

Cantua quercifolia

This low shrub grows in sunny thickets in the mountains of Ecuador and northern Peru. Dr. Calaway Dodson found it growing at an elevation between 6000 and 7200 feet near Loja, Ecuador, which is a higher altitudinal zone than that inhabited by the related *C. pyrifolia* in the same area. It flowers heavily during the rainy season from April to May and intermittently through the rest of the year.

The white trumpet-shaped flowers stand erect in terminal clusters. They have a broad corolla tube 2 cm. long, which contains copious quantities of nectar, and long exserted stamens and style.

Dr. Dodson has observed hummingbirds, hawkmoths, wasps, and bees visiting the flowers for nectar. The smaller insects are more or less ineffective as pollinators. The hummingbirds probably bring about a great deal of pollination of the Cantua flowers. The hawkmoths and large wasps probably effect some or much pollination too.

FLORAL MECHANISM. The plants flower most heavily during the rainy season from April to May in the hills near Loja, Ecuador, according to Dr. Calaway Dodson (personal communication), but continue to flower intermittently throughout the rest of the year. The white flowers are grouped in terminal clusters in the upper part of the plant and stand erect. They have a faint sweet odor.

The white corolla consists of a stout tube (2 cm. long and 4 or 5 mm. in diameter) and a flaring limb. Nectar is secreted copiously at times, and if allowed to accumulate may fill up the corolla tube almost to the brim. The stamens and style are exserted about 2 cm. above the entrance to the tube. The five bright yellow anthers and the stigma are all on the same level.

ANIMAL VISITORS. Dr. Calaway Dodson observed several kinds of animals visiting the flowers in the population near Loja, Ecuador. An unidentified medium-sized hummingbird fed at intervals on the flowers throughout the day, hovering and probing for nectar. Unidentified hawkmoths visited the flowers for nectar occasionally by day and more frequently at night. Small bees, a vespid wasp (*Polybia occidentales*), and a large scoliid wasp (*Campsomeris ephippium*) landed on the flowers and poked into the tube for nectar during the daytime.

Some of the insect visitors, particularly the smaller bees and wasps, do not normally contact the long exserted stamens and style in their approach to the flowers, and can be excluded as effective pollinators. The large scoliid wasp often does brush against the anthers or stigma in making a landing, though it may also approach the flowers from the side, and so it is a partially effective pollinator according to Dr. Dodson. The hawkmoths probably transfer some pollen. The most effective pollinating agent among the observed visitors may well be the hummingbirds, which feed frequently and undoubtedly carry much pollen on their head and chest.

Cantua pyrifolia   *Fig. 38 B*

This evergreen Andean shrub has waxy white, erect, tubular flowers with long exserted stamens and style. The corolla tube is somewhat shorter (1.5 cm.) and narrower than in *C. quercifolia*. Plants grown in Claremont have completely protandrous but self-compatible flowers.

Cantua candelilla   *Plate II A; Figs. 39, 40*

Cantua candelilla is a shrub which grows among columnar cacti in an open dry vegetation on the west slope of the Andes in southern Peru. The reddish, odorless, trumpet-shaped flowers hang down loosely at the ends of the arching branches. These flowers are among the largest in the Phlox family with a tube 5.5 cm. long. The nectar in these floral tubes could be reached successfully by long-billed Andean hummingbirds, as for example Patagona and Coeligena, which would normally contact the anthers and stigma in feeding. Such humming-birds may be considered as probable pollinators of this plant.

FLORAL MECHANISM. The following notes are based on plants grown in Claremont from seeds collected by Mr. H. Johnson near Tarata, Arequipa, Peru.

The large trumpet-shaped flowers hang down in loose cymes at the ends of the arching branches. The corolla varies in color from reddish-orange to yellow in different individuals. The corolla tube is 5.5 cm. long and 6 or 7 mm. wide at the entrance. The short lobes form a funnel-like limb 2 cm. broad. The stamens and style are exserted beyond the limb on the lower side of the flower. The flowers are incompletely protandrous.

Nectar is secreted in copious amounts and may accumulate to a depth of 1 cm. in the corolla tube. It is held in position in the pendant flowers partly by its viscosity and partly by capillary attraction to the stamen bases and corolla wall. The distance from the entrance of the corolla to the nectar is 4.0 to 4.5 cm.

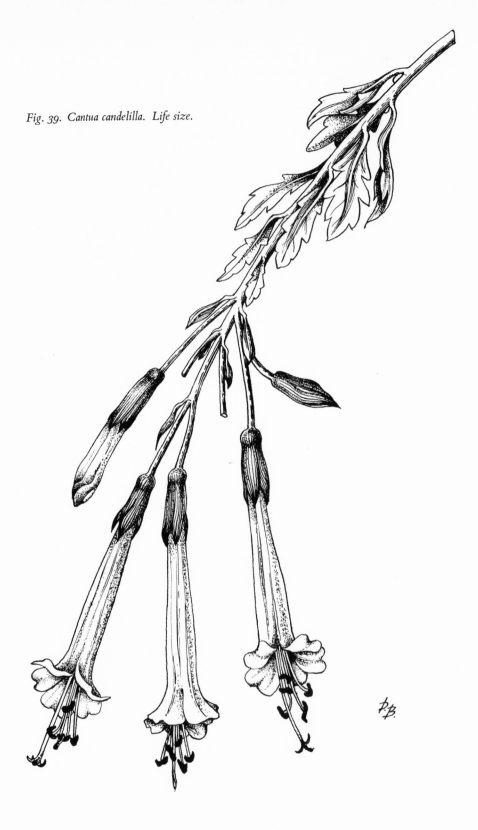

Fig. 40. Size relations between western American and Andean hummingbirds and hummingbird flowers. Above: Ipomopsis aggregata and Selasphorus rufus. Below: Cantua candelilla and Patagona gigas. All life size.

BREEDING SYSTEM. The Tarata strain is self-compatible.

MODE OF POLLINATION. The distinctive features of the flowers of *Cantua candelilla,* their trumpet-like form, hanging position, and reddish color, are the characters of hummingbird flowers. Their large size, larger than any other bird flowers in the family, suggest that they are fitted for some particularly long-billed hummingbirds. It is well known that the Andes are inhabited by a diversity of hummingbirds, including some long-billed species.

The senior author went through the collection of Peruvian hummingbirds in the American Museum of Natural History some years ago in a search for species with bills long enough to fully probe the floral tube of *Cantua candelilla.* Three species which are apparently common in the area of *Cantua candelilla* have long bills as follows: *Patagona gigas peruviana* (4.0 cm. long bill), *Coeligena torquata* (3.5 to 4.0 cm.), *Coeligena violifer* (3.5 cm.). A fourth species, *Ensifera ensifera ensifera,* has an extremely long bill (6.0 to 10.5 cm.).

The bills of dead bird specimens belonging to these species were inserted manually into live Cantua flowers. In long-billed individuals of Ensifera the head and chest would stand off too far from the stamens to pick up much pollen. The 4 cm. long, stout bill of Patagona, on the other hand, fits perfectly in the floral tube up to the nectar-containing region, and the chin of the bird then contacts the anthers or stigma. *Coeligena torquata* could also reach the nectar and pollinate the flowers.

Since this went to press, Dr. Richard Straw informs us that he observed *Patagona gigas* visiting the flowers of *Cantua candelilla* in the mountains above Arequipa, Peru, on a recent expedition.

Cantua buxifolia

Cantua buxifolia, which was widely grown by the Incas, and is now found mainly if not exclusively in cultivation or as an escape in Peru, is related to *C. candelilla* and has similar large, pendant, trumpet-shaped flowers. Loew suggested bird pollination for *C. buxifolia* (Knuth and Loew, 1905, 58). As with *C. candelilla,* we would expect long-billed hummingbirds such as Patagona and Coeligena to be able to work the floral mechanism most successfully.

HUTHIA   *Plate II B*

Huthia is a Peruvian genus of small shrubs related to Cantua. The two species differ markedly in floral characters. *Huthia coerulea* has horizontal, violet flowers with an extended lower lip, expanded throat, and tube 1.0 to 1.3 cm. long. Its flowers are suggestive of bee pollination. In *Huthia longiflora* the flowers are large and trumpet-shaped, with a tube 4 cm. long and about 5 mm. wide at the orifice. These flowers seem to be best fitted for visits by long-billed hummingbirds.

COBAEA SECTION COBAEA

Cobaea is a poorly known genus of woody-based climbing vines which occurs in the tropics from Mexico through Central America to the Andes. The plants grow in tropical forest and woods where they climb by means of tendrils. They expose their large flowers in the open air outside the foliage on long stout peduncles. The form and position of the flowers differs between the different sections of Cobaea.

In the section Rosenbergia the flowers are pendant with long dangling ribbon-like petals. In the section Cobaea the flowers are nodding, rather than fully pendant, and are campanulate with a large cup-shaped throat and short lobes. This is the largest and most widespread section of the genus, containing 13 species ranging from Mexico to the Peruvian Andes.

Cobaea scandens   *Figs.* 41, 42

The natural area of distribution of *Cobaea scandens* lies in the montane tropical zone of south-central Mexico. It also grows as an escape from cultivation there and elsewhere in the tropics. It was introduced into cultivation in Europe in 1787 and has since been grown fairly widely there and in the United States as an ornamental vine. *Cobaea scandens* is consequently well known to botanists and horticulturists living in the temperate zone, and various aspects of its morphology and growth have been investigated in some detail, but unfortunately the ecology of the plants in their natural environment is virtually unknown.

The large campanulate flowers are borne at the end of a long stout peduncle which holds them in a slightly nodding position in the open air outside the foliage. The corolla when it first opens is greenish-yellow and has a cabbage-like odor; then as it reaches full bloom it turns purple and ceases to give off the sour odor. The large bell-shaped throat forms a chamber 3 cm. deep and 2 to 3 cm. wide at the entrance. Below the floral chamber is a small nectar chamber roofed over by a covering of hairs.

The stamens and style are exserted well beyond the open entrance to the floral chamber and on its lower side. The flowers are completely protandrous. The plants are self-incompatible (Behrens, 1880).

The identity of the pollinating agent of *Cobaea scandens* in its natural habitat has been the subject of considerable and long-standing speculation. Among older

Fig. 41. Cobaea scandens. Life size.
(Redrawn from Scholtz, 1893.)

authors, Behrens (1880) suggested bumblebees, Scholtz (1893) hawkmoths, and Knuth (1909, in the original German edition of 1899) hummingbirds. These suggestions can be set aside. The floral chamber is much wider than the bodies of bees or the bills of hummingbirds or hawkmoths, with the result that these animals can (and in gardens sometimes do) enter the flowers without contacting the anthers or stigma. Vogel (1958) has put forward the far more plausible suggestion that *Cobaea scandens* is a bat flower.

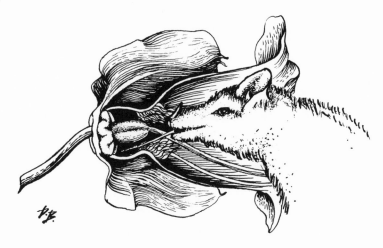

Fig. 42. Flower of Cobaea scandens and head of Leptonycteris nivalis, taken as a representative glossophagine bat, to show size relations. Life size.

Vogel was led to this suggestion by the finding of bat claw marks on the flowers of a related species, *Cobaea trianaei*, as will be described later. He also noted that the flowers of the *Cobaea scandens* group are generally similar in their large size, campanulate shape, stout support, muddy hue, and sour odor to other known bat flowers, such as *Crescentia cujete* (Bignoniaceae), *Kigelia aethiopica* (Bignoniaceae), *Symbolanthus latifolius* (Gentianaceae), and *Campanea grandiflora* (Gesneriaceae).

The flower-visiting bats of the American tropics belong to the subfamily Glossophaginae of the vegetarian family Phyllostomidae. These bats depend mainly on nectar and pollen for food which they obtain at night. They characteristically settle down in the entrance to a flower, and force their head and shoulders into the floral chamber to feed, while clinging to the outer wall of the corolla with their wing claws. They have a slender muzzle and long narrow extensile tongue with a brush-like tip which can extract nectar from a buried nectar chamber (see Porsch, 1934–1935; Pijl, 1936, 1956; Allen, 1939; Jaeger, 1954; Vogel, 1958).

Through the courtesy of Dr. E. L. Cockrum of the University of Arizona we obtained some pickled specimens of the glossophagine bat, *Leptonycteris nivalis*, which is typical in size and shape of its group, and inserted the specimens into live flowers of *Cobaea scandens*. The head of the bat fits very well into the floral chamber of the Cobaea and its tongue is in a good position to probe through the constricted hairy base of the chamber into the nectar reservoir. The bat specimens had Cobaea pollen adhering to the fur on their breast and chin after a simulated visitation. The same body parts contact the stigma in the course of a visit to a flower in the female stage.

FLORAL MECHANISM. Cobaea scandens has been extensively cultivated in Europe, the United States, and elsewhere, and numerous botanists have had an opportunity to study its floral mechanism. The following account is based on the older discussions by Behrens (1880), Scholtz (1893), and Kerner (1895), and on original observations of plants grown in Claremont.

The large bell-shaped flowers are a muddy purple at maturity and are borne singly at the ends of stout peduncles 12 to 22 cm. long. The cup-shaped floral chamber is 3 cm. deep from the entrance to the base, and the elliptical entrance is 2 to 3 cm. wide. The short corolla lobes spread to form a limb about 5 cm. wide. Below the floral chamber is a tube containing the nectar. The small nectar chamber is separated from the large floral chamber by a constriction and a dense tuft of hairs.

The stamens lie on the lower side of the corolla and are exserted well beyond the limb. Near their distal ends the filaments bend sharply upward to position the long versatile anthers about 1 cm. beyond the entrance to the floral chamber and on its lower side. Some of the anthers normally dehisce and expose the ripe pollen before the other anthers become ripe. The time lag between the early and late anthers varies from a few hours to a full day. As the anthers pass maturity, the filaments bend backward and retract. The style, heretofore not fully developed, now elongates and occupies the position previously occupied by the anthers, and the stigma lobes open out and become receptive.

The protandrous development of the sex organs is part of a larger developmental sequence involving growth movements of the peduncle, the formation of color pigments in the corolla wall, and other changes, which we may now consider.

The peduncle holds the flower bud in an erect position. The tip of the peduncle bends downward slightly as the flower opens. At this stage the peduncle is ascending through most of its length, bringing the flower well outside the zone of foliage, but is bent downward at the tip to place the flower in a nodding position. After the flower passes maturity the peduncle bends down further so that the fruit develops in the shelter of the foliage.

The corolla as it first opens is greenish-yellow and gives off a sour cabbage-like odor. During the first day that the corolla is open it starts to turn purple. By the time the anthers dehisce the corolla is purple and has lost its disagreeable odor. An individual flower remains in full bloom for three, four, or five days (in Claremont), being in the staminate condition during the first few days and in the stigmatic condition during part of the last day.

BREEDING SYSTEM. Behrens (1880) found that *Cobaea scandens* is self-incompatible, and we can confirm this finding from tests made on plants grown in Claremont.

ANIMAL VISITORS. Flower visitors have been seen only on plants cultivated in Germany or California far outside their natural range. Such observations are of little value. Our attempts to fill this gap in our information by field studies in the Mexican tropics were unsuccessful.

Behrens (1880) and Scholtz (1893) observed *Bombus muscorum* repeatedly visiting the flowers in Germany and sometimes pollinating them. Scholtz (1893) also recorded visits by *Vespa germanica* (Vespidae), *Eristalis tenax* (Syrphidae), and *Macroglossa stellatarum* (Sphingidae).

In Claremont, California, the flowers are ignored by carpenter bees (*Xylocopa*), which nest near them and fly by them without exhibiting any interest.

However, Anna hummingbirds (*Calypte anna*) regularly feed on the Cobaea flowers in a garden in Claremont. In hovering to probe for nectar their chest comes into contact with the anthers. The hummingbirds bring about some pollination but not much. Six large plants produced numerous flowers which were fed on frequently by the hummingbirds. These feeding visits led to the formation of a total of six capsules during a two-month period.

Cobaea trianaei   Fig. 43

This species of the mountain forests of Colombia has campanulate flowers held out on long stout peduncles. The flowers bloom principally at night and give off a pungent odor (Vogel, 1958). The corolla is greenish-yellow with muddy violet veins. The floral chamber has the same campanulate shape as in *Cobaea scandens*, but is slightly smaller in size, and the stamens and style have the same exserted position and protandrous order of development. An individual flower usually blooms through several nights before the corolla falls to the ground (Vogel, 1958).

Vogel (1958) observed *Cobaea trianaei* in a natural habitat in the Andes of Colombia. The plants were growing in moist forest at an elevation of 6000 feet near Tequendama Falls in Cundinamarca province. Here the plants climbed to heights of 45 feet in the forest canopy, where they flowered, and they were in full bloom in November.

On all the fallen corollas Vogel found the claw marks of bats. These marks usually occurred on the outer lateral walls of the corolla throats where bats customarily cling by their wings in feeding.

Vogel (1958) was able to actually observe glossophagine bats visiting the flowers of several other plant species, namely *Marcgravia rectiflora* (Marcgraviaceae),

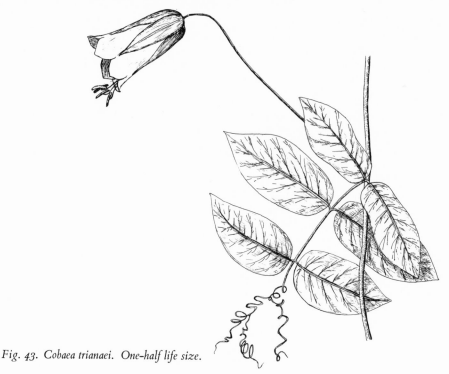

Fig. 43. Cobaea trianaei. One-half life size.

Purpurella grossa (Melastomataceae), and *Symbolanthus latifolius* (Gentianaceae), at other localities in the Colombian Andes. In the immediate vicinity of Tequendama where *Cobaea trianaei* was growing he found further indirect evidence of bat activity in the form of claw marks on the flowers of *Mucuna mutisiana* (Papilionaceae).

The bats probably cannot enter the Cobaea flowers in the manner indicated by the position of their claw marks on the outer corolla wall without brushing against the anthers or stigma.

Cobaea lutea

The floral mechanism and breeding system of this Guatemalan species have been studied by Ross (1898) on the basis of plants grown in Germany. The flowers are campanulate, greenish-yellow, and vespertine or nocturnal in their periodicity. The exserted stamens and style are mature during the same night and come into contact with one another automatically. The plants are self-compatible and are autogamous to some extent.

FLORAL MECHANISM. Ross (1898) has described the floral mechanism on the basis of plants grown in Germany (under the synonym, "*Cobaea macrostemma*").

The campanulate flowers, borne as in the other species terminally on long peduncles and oriented outward and slightly downward, are greenish-yellow and have no odor. The corolla throat is somewhat smaller but the limb is longer than in *Cobaea scandens*. The nectar chamber lies below the corolla throat and is separated from it by a constriction. The flow of nectar commences in late afternoon or dusk.

The stamens are well exserted, unequal in length, and spread apart widely in front of the entrance to the corolla throat. The anthers dehisce in the late afternoon and remain in a ripe condition through the night.

Several hours after anther dehiscence, at dusk or in the early evening of the same day, the style bends upward at the tip and the stigma opens. It too remains ripe through the night. Ross states that the style makes a slow oscillating motion from side to side which occasionally brings the stigma into contact with one or another of the anthers.

On the following morning the stamen filaments coil up, and the style, once pollinated, bends downward.

Artificial pollinations made by Ross in the first evening of full bloom led to a greater set of fruits than pollinations carried out the following morning.

BREEDING SYSTEM. Ross (1898) found that artificially self-pollinated and cross-pollinated flowers were equally fruitful. The number, soundness, and germination of seeds resulting from selfing is on a par with that resulting from crossing. It thus appears that *Cobaea lutea*, unlike other known species of the genus, is self-compatible.

We have seen above that the floral mechanism permits some self-pollination to take place automatically. Ross observed that the plants grown in a greenhouse in Germany set fruits freely in the absence of insect visits.

COBAEA SECTION ROSENBERGIA

This section contains five known species in the mountains from Costa Rica to Ecuador. The corolla has a small or poorly developed throat region, as compared with the section Cobaea, and very long corolla lobes.

Cobaea penduliflora   *Fig.* 44

This species inhabits moist forest at an elevation of 6000 feet in the mountains back of Caracas, Venezuela, and has been studied by Ernst (1880) in a garden in Caracas. The pendant flowers consist of long green ribbon-like corolla lobes (7 to 10 cm. long) dangling from the basal nectar-containing cup, a whorl of

Fig. 44. Cobaea penduliflora. Life size.
(Redrawn with modifications from Hooker, 1869.)

stamens on long reddish-purple filaments, and a still longer green style. The flowers are nocturnal in their period of nectar flow and anther and stigma ripening. The plants are self-incompatible.

At night the flowers are visited by hawkmoths (Cocytius, Phlegethontius, Xylophanes), which hover near the anthers and stigma and insert their proboscis into the nectar cup. In this position the wing tips pick up and carry the sticky pollen and deliver it to the stigma of another flower. Flowers not visited by the moths were sterile.

FLORAL MECHANISM. Ernst (1880) has described the behavior of plants grown in a garden in Caracas, Venezuela, close to the natural habitat of the species in the neighboring mountains. The following account is drawn mainly from Ernst and partly from Hooker (1869).

The flowers hang down at the ends of the long peduncles. The corolla and style are green, the stamen filaments reddish-purple, and the anthers yellow. Ernst could detect no odor.

The corolla has a small cup-shaped base, which contains the nectar, and five long, linear, wavy-margined lobes which dangle like streamers below the cup. The lobes are 7 to 10 cm. long and the cup is about 2.5 cm. deep and 1 cm. wide.

The stamens on filaments 7 cm. long hang down and spread apart so that the anthers stand in a whorl about 15 cm. in diameter around the dangling corolla lobes. The style exceeds the stamens in length. Ernst (1880) states that the stigma assumes "a central position," whereas Hooker (1869) depicts it as projecting outward to one side of the flower, and perhaps its exact position varies. The stigma is well separated spatially from the anthers.

Nectar is secreted in copious quantities by the disk at the base of the ovary, accumulates in the cup-like base of the corolla, and is retained there by a covering of hairs.

The flowers are nocturnal. The nectar starts to flow in late afternoon or early evening. The anthers dehisce and the stigma becomes receptive at sundown. A given flower blooms during a single night. The following morning nectar secretion ceases, the stamens coil up, and the stigma becomes unreceptive to pollen. Ernst artificially cross-pollinated several flowers in the evening and several others the next morning; the former set fruit but the latter did not.

After the flower has passed blooming, the peduncle withdraws slowly into the dense foliage, and the fruit develops in the shade of the leaves.

Although each flower blooms only once, a constant succession of flowers appears night after night during the blooming season.

BREEDING SYSTEM. Twelve flowers self-pollinated by Ernst (1880) produced only one capsule, and that one could have been due to accidental contamination, whereas nine flowers cross-pollinated under the same conditions set fruits, from which it can be concluded that these plants are self-incompatible.

INSECT VISITORS. In the garden in Caracas the flowers were visited in the evening by three kinds of large-bodied long-tongued hawkmoths, viz., Cocytius, Phlegethontius, and Xylophanes. (The names given by Ernst were Amphonyx, Diludia, and Chaerocampa, respectively, and we have listed their probable equivalents in present-day nomenclature.)

The hawkmoths hovered with their body close to the stigma and their proboscis stretched out to the nectar cup. The beating of their wings caused the wing tips to become covered with the sticky yellow pollen. Ernst observed that the moths in visiting other flowers touched the stigma with their wings and left pollen on it. All such flowers set fruit. Some flowers which were not visited by the moths, on the other hand, remained sterile.

10

SURVEY OF
POLLINATION SYSTEMS

The purpose of this chapter is to provide a brief summary of the modes of pollination in the Phlox family.

POLEMONIUM TRIBE

Polemonium

Most of the species of Polemonium are cross-fertilizing perennial herbs with fragrant, blue, campanulate or broad funnelform flowers. The stamens and style stand on the lower side of the broad corolla throat and the nectar is concealed in a short tube below the throat. In the species of the lowland temperate zone, such as *P. caeruleum*, *P. foliosissimum*, and *P. reptans*, the flowers are pollinated chiefly by bumblebees and other large bees, which perch in the broad corolla throat while feeding and carry pollen on their venter (Fig. 1).

The species inhabiting high mountains or arctic regions, such as *P. delicatum*, *P. californicum*, *P. viscosum*, and *P. boreale*, are pollinated by various scavenger and nectar-feeding flies (Anthomyiidae, Muscidae, Calliphoridae, Empididae, Syrphidae), as well as by bees. The skunk-like odor of the herbage which is characteristic of most alpine and arctic species of Polemonium may attract the flies to the plants.

Both hummingbirds and bumblebees pollinate the long tubular-funnelform flowers of *Polemonium confertum* in the Rocky Mts. Two other species, *P. brandegei* and *P. pauciflorum*, have trumpet-shaped flowers which seem to be fittted specially for hummingbirds, though supporting observational evidence is lacking (Plate II C).

The annual species, *P. micranthum*, has reduced white flowers which produce seeds exclusively or nearly exclusively by self-pollination (Fig. 3). A less extreme development of facultative autogamy is found in *P. mexicanum*, *P. pulcherrimum*, and *P. pauciflorum*.

Collomia, Allophyllum, and Gymnosteris

The genus Collomia, which is related to but more advanced than Polemonium, has smaller, more slender funnelform flowers. These flowers are known in some cases to be pollinated by beeflies (Bombyliidae). Several annual species of Collomia with quite small flowers (i.e., *C. linearis*, *C. heterophylla*) are predominantly autogamous. Cleistogamous flowers are found in some strains of the otherwise showy-flowered *C. grandiflora* (Fig. 4).

The related minor genus Allophyllum contains five species of annual herbs. The blue funnelform flowers of *A. glutinosum* are pollinated by small bees. The salverform flowers of *A. divaricatum* are visited and pollinated by beeflies, principally *Bombylius lancifer*, which insert their needle-like proboscis into the long slender corolla tube for nectar (Fig. 5). The remaining three species are predominantly autogamous.

Gymnosteris, another minor genus related to Collomia, contains two species of annuals with reduced flowers. Autogamy is known in one of these species (*G. parvula*) and inferred in the other (Fig. 6).

Phlox and Microsteris

The species of Phlox are with few exceptions cross-fertilizing perennial herbs or subshrubs with salverform flowers pollinated by Lepidoptera. The salverform corolla has a long slender nectar tube and a broad spreading colorful limb, often with an eye spot around the orifice, and it generally gives off a sweet fragrance. Various kinds of Lepidoptera probe for nectar in these flowers, from either a perching or hovering position depending on their habits, and carry pollen on their proboscis or face.

Many herbaceous species of Phlox with corolla tubes 1 to 2 cm. long and broad inflorescences (i.e., *P. glaberrima*, *P. pilosa*, *P. paniculata*) are pollinated by perching butterflies with tongues similar in length to the floral tubes (Fig. 7). *Phlox stansburyi* and *P. dolichantha* with corolla tubes 3 to 4 cm. long and loose inflorescences appear to be fitted for pollination by long-tongued hovering hawkmoths (Fig. 8). A transitional condition is found in *P. drummondii* which is pollinated by both butterflies and hawkmoths.

The cespitose western Phloxes with corolla tubes 1 cm. long borne erect at the ends of matted branches (i.e., *P. caespitosa*, *P. diffusa*) are pollinated by noctuid moths with tongues of similar length and perching feeding habits (Fig. 9).

The small-flowered annual, *Microsteris gracilis*, which is close to Phlox and probably derived from it, is predominantly autogamous, though occasionally visited and cross-pollinated by insects.

GILIA TRIBE

Gilia

Each of the five sections of Gilia contains some cross-fertilizing species with broad funnelform or campanulate, blue bee flowers. Bumblebees are the usual pollinators of *Gilia pinnatifida* in the Rocky Mts. The large black anthophorine bee Tetralonia pollinates G. *latiflora* in the Mojave Desert (Plate I E). Small halictid, megachilid, and andrenid bees are the common pollinators of such California species as *Gilia tricolor* and *G. ochroleuca bizonata* (Fig. 11).

In three of the five sections there are species or races with slender funnelform flowers which are visited and pollinated by beeflies (Bombyliidae). *Gilia splendens splendens*, for example, is pollinated mainly by *Bombylius lancifer*, the medium-long proboscis of which is fitted to reach the nectar in the slender corolla tube (Fig. 15). Mixed pollination by both beeflies and bees represents a transitional condition found in various species (i.e., G. *tricolor*, G. *leptantha pinetorum*).

An extreme development for pollination by long-tongued flies is found in certain localized races of *Gilia splendens*, G. *leptantha*, and G. *cana*, which possess very long slender corolla tubes (1.5 to 2.0 cm. long). The fly, *Eulonchus smaragdinus* (Cyrtidae), with a thin straight proboscis of similar length to the nectar tubes, feeds on and pollinates the long-tubed race of G. *splendens* (Fig. 15).

Hummingbird pollination occurs in another local race of *Gilia splendens* with broad tubular flowers, and is expected from the flower form in G. *subnuda* (Fig. 15).

The majority of species of Gilia are reduced annuals with small self-pollinating flowers. Autogamy is found in every section of the genus and in an estimated 38 species. This condition is particularly common among the species which have colonized the western American deserts (i.e., G. *minor*, Fig. 14), the high mountains (G. *capillaris*), and southern South America (G. *laciniata*).

Ipomopsis

Ipomopsis contains 27 species grouped into three sections: Phloganthea, Ipomopsis, and Microgilia. For a genus of moderate size, it exhibits a wide diversity of flower forms and modes of pollination.

Blue bee flowers are found in the section Phloganthea, as exemplified by *I. multiflora* (Plate I K).

Red trumpet-shaped hummingbird flowers are developed in various species of the sections Phloganthea and Ipomopsis, thus in *I. tenuifolia* and *I. aggregata* (Plate II D, E; Fig. 16).

The section Ipomopsis also includes species with long-tubed violet or whitish hawkmoth flowers (Plate III C). *I. tenuituba* and *I. candida* with floral tubes ranging from 3.0 to 4.5 cm. long are visited and pollinated by Celerio and Sphinx with tongues of similar length (Fig. 17). *Ipomopsis macombii* in the same section has violet flowers with a shorter nectar tube (1.5 cm. long), which are fed on and pollinated by skippers and butterflies (i.e., Ochlodes and Papilio) (Fig. 18).

In the third section, Microgilia, *Ipomopsis congesta* has small whitish flowers clustered in capitate heads. Melyrid beetles commonly visit and pollinate these flowers (Fig. 19). The flowers of the related *I. spicata* are small and yellowish and give off a cloying nitrogenous odor, which attracts scavenger flies such as the tachinid Athanatus (Fig. 20).

Eight annual species in the section Microgilia, finally, have small inconspicuous flowers which are either known or inferred to be predominantly autogamous (i.e., *I. pumila*, Fig. 21).

Eriastrum

Unlike Ipomopsis, to which it is related, Eriastrum is fairly homogeneous in floral characters. The corolla is funnelform, usually blue but yellow in one species, and the flowers are often grouped in small heads (Plate I H, I, J). Several species are cross-pollinated mainly by bees (i.e., *E. densifolium*, *E. sapphirinum*). A Mojave Desert species, *E. eremicum*, is pollinated by the beefly Pantarbes. Some desert species with small flowers (i.e., *E. wilcoxii*) are probably predominantly autogamous.

Langloisia

This minor genus of small desert annuals related to Eriastrum contains four species. Two of these have rather bizarre flowers which are pollinated by beeflies.

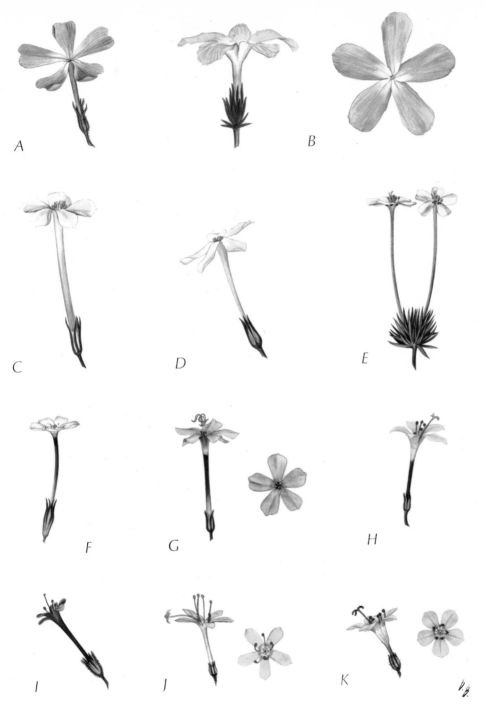

Plate III. Lepidoptera flowers (A–D) and long-tongued Fly flowers (E–K). All life size.

(A) Phlox divaricata.   (B) Leptodactylon californicum.
(C) Ipomopsis longiflora.   (D) Phlox stansburyi.   (E) Linanthus parviflorus.
(F) Linanthus montanus.   (G) Gilia cana speciosa.   (H) Gilia splendens (San Gabriel race).
(I) Ailophyllum divaricatum (Sierran race).   (J) Gilia leptantha leptantha.   (K) Gilia cana triceps.

In *Langloisia punctata* the flowers are open and bell-shaped with a whitish background color decorated by numerous purple dots. The flowers of *L. matthewsii* are horizontal and distinctly bilabiate, an unusual condition in the Polemoniaceae, and are pollinated by the beefly Pantarbes and to a lesser extent by bees (Fig. 22). The two small-flowered species are probably autogamous.

Navarretia

The annual genus Navarretia, like Eriastrum to which it is related, is characterized by funnelform flowers which are usually blue but sometimes yellow, pink, or white (Fig. 23). These flowers are pollinated by small bees and/or beeflies. A highly specialized group of species inhabiting vernal pools and possessing tiny flowers with included sex organs is probably autogamous.

Leptodactylon

The cross-fertilizing shrub, *Leptodactylon californicum*, has bright pink, fragrant, salverform flowers. Various Lepidoptera such as *Papilio rutulus*, *Hemaris senta*, and others visit the flowers for nectar and carry the pollen on their proboscis (Fig. 24).

Linanthus

Linanthus contains 38 species of herbaceous plants, mostly annuals, which exhibit a wide diversity of flower forms and pollination systems. The six sections of this heterogeneous genus fall into two main branches as follows: (1) Siphonella, Pacificus, Leptosiphon, Dactylophyllum; and (2) Dianthoides, Linanthus.

Broad funnelform bee flowers with a cup-shaped throat are found in three of these sections belonging to both phylogenetic branches. In *Linanthus grandiflorus* (section Pacificus) the corolla throat is large and the pollinators are large bumblebees (Fig. 25). *Linanthus liniflorus* (section Dactylophyllum), on the other hand, has small cup-shaped flowers borne on long slender stalks. These flowers are pollinated by small bees (i.e., Dufourea, Chelostoma, Ashmeadiella) (Fig. 27).

The section Leptosiphon is distinguished by salverform flowers with long slender nectar tubes. These flowers are adapted for pollination by long-tongued flies. The races of *Linanthus parviflorus* with corolla tubes of medium length are pollinated by beeflies, especially *Bombylius lancifer*. A race of *L. androsaceus* with an extremely long corolla tube (2.5 to 3.0 cm.) is pollinated by the cyrtid fly, *Eulonchus smaragdinus*, with a needle-like proboscis 2 cm. long (Fig. 26).

The tufted desert annual, *L. parryae* (section Dianthoides), with cup-shaped flowers, is pollinated by the melyrid beetle Trichochrous (Fig. 28). *Linanthus dichotomus* in the related section Linanthus has fragrant white vespertine moth flowers (Fig. 29).

Some species belonging to five of the six sections have small inconspicuous flowers which are probably predominantly autogamous.

BONPLANDIA TRIBE

Bonplandia

The leafy shrub, *Bonplandia geminiflora*, of the Mexican tropics has blue bilabiate flowers with a long corolla tube and an extended lower lip (Plate I A, Fig. 38 A). These flowers seem to be well suited for visitations by long-tongued tropical bees, though field observations are needed.

Loeselia

Loeselia is poorly known in the field although several species have been grown and studied in the garden during this investigation. In some species the flowers are blue or yellow and bilabiate, suggesting bee pollination. *Loeselia mexicana* has red trumpet-shaped hummingbird flowers (Plate II F, Fig. 38 C).

CANTUA TRIBE

Cantua

This Andean genus of shrubs and small trees contains four species with trumpet-shaped flowers. The erect white tubular flowers of *C. quercifolia* are 2 cm. long and are visited and pollinated by hummingbirds and hawkmoths, perhaps most effectively by the former. The pendant reddish flowers of *C. candelilla* and *C. buxifolia* are very large (tube 5.5 cm. long), suggesting that they are adapted for pollinating visits by long-billed hummingbirds like *Patagona gigas* (Plate II A, Fig. 39).

Huthia

The flowers of the Peruvian shrub, *Huthia longiflora*, are large and trumpet-shaped, like those of *Cantua candelilla* in the same general area, and may well be pollinated by long-billed hummingbirds (Plate II B). A second species, *H. coerulea*, has smaller violet flowers suggestive of bee pollination.

COBAEA TRIBE

Cobaea

This genus of tropical vines has large nocturnal flowers borne singly on long stout stalks. There are two basically different types of floral mechanism in the genus.

In *Cobaea scandens*, *C. trianaei*, and related species, the flowers are campanulate. They have a large cup-shaped throat with nectar concealed in a chamber at the base and are often yellowish or greenish with a sour or pungent odor (Figs. 41, 43). The flowers of *C. trianaei* and presumably of other similar species are visited by nectar-eating glossophagine bats.

In *Cobaea penduliflora* and its relatives, on the other hand, the flowers are pendant with long dangling ribbon-like petals and long stamens and style. These flowers are visited by long-tongued hovering hawkmoths (Cocytius, Phlegethontius, Xylophanes) (Fig. 44).

THE POLLINATION SPECTRUM

Ten main modes of pollination are found within the family Polemoniaceae. These are cross-pollination by bees, long-tongued flies, scavenger flies, hummingbirds, butterflies, hawkmoths, noctuid moths, beetles, and bats; and self-pollination by autogamy or (very rarely) cleistogamy.

It is of interest to estimate the relative frequency of the different pollination systems in the family. We have pollination records for 122 of the 327 species of Polemoniaceae. The probable mode of pollination of some 80 additional species can be predicted fairly reliably from what is known about their relatives.

In classifying the species of Polemoniaceae with respect to pollination system, only the principal mode of pollination has been taken into account, and casual or exceptional pollen-carrying agents have been ignored. In many cases two classes of agents participate more or less equally in the pollination of a given species, where one race is pollinated chiefly by bees and another chiefly by beeflies, or where the same race is pollinated effectively by both bees and beeflies. Such situations have been handled for statistical purposes by scoring the species as one-half bee flowers and one-half beefly flowers.

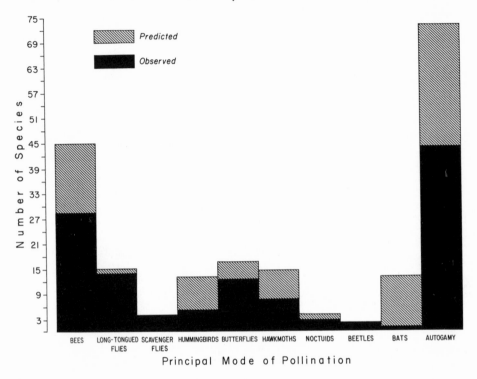

Fig. 45. Frequency distribution of different modes of pollination in the Phlox family.

The frequency distribution of the 10 different pollination systems in the Phlox family is shown by the bar graph in Fig. 45. This graph shows that bee pollination is the most common method of cross-pollination. Some other modes of cross-pollination, as by long-tongued flies, hummingbirds, and Lepidoptera, are also well represented in the family. Cross-pollination by beetles and scavenger flies is rare. Autogamy, on the other hand, is quite common, being found in more species than any single method of cross-pollination.

II

ADAPTIVE RADIATION

In this final chapter we shall consider the evolutionary significance of the observations reported in the preceding chapters. The diversity of flower forms and pollinating agents in the Phlox family is a fact of observation. Our thesis is that this diversity is the result of a process of adaptive radiation in the method of pollination.

In developing this thesis we will take up three main subsumptive problems in turn. First, is the postulated radiation of flower forms in fact adaptive? Second, what is the probable historical course of the radiation? And finally, how, considered as a process, has the radiation come about?

COADAPTATION

The premise that flowers with different floral mechanisms are adapted to different kinds of pollinating agents, which was accepted by the classical floral biologists, is viewed with skepticism or rejected altogether by some modern students. The diversity of opinion among pollination students today as regards evolutionary concepts is such that almost any point of view, and especially the one presented here, is bound to be controversial. An exposition and critique of the various evolutionary viewpoints, many of which are obsolete, would, however, take us too far afield. Rather than enter into the task of literary criticism here, therefore, we propose to summarize, out of the background of experience gained during the present study, the considerations which have led to an interpretation closely approximating the classical one. For recent critical reviews with references the reader may be referred to Takhtajan (1959) and Pijl (1960, 1961); for recent discussions of the problems involved see Baker (1961, 1963).

In *Ipomopsis candida* and *I. thurberi* the flowers are long and salverform, with nectar tubes 3 to 4 cm. long, and are borne in loose racemes. They are nocturnal in periodicity, the nectar flow and fragrance being greatest at night, and the whitish or violet colors being visible at night.

These flowers are ignored by most of the flower-visiting animals around them during the daytime. An occasional large bee will attempt to collect pollen from them, but the flowering branch bends down precariously under the weight of such bees, and they soon give up the attempt. Small halictid bees can and do collect pollen with greater success and greater frequency by day, and bring about some pollination in the course of their visits.

At night long-tongued hawkmoths such as *Celerio lineata* and Sphinx feed on the nectar of the flowers from their characteristic hovering position. They fly rapidly and frequently from plant to plant, carrying pollen on their proboscis and pollinating numerous flowers with much effectiveness. On warm summer nights at the height of the blooming season the hawkmoths may visit an Ipomopsis population in considerable numbers.

The *Ipomopsis candida* and *I. thurberi* flowers conform in size, form, orientation, color, and periodicity to the structural characteristics and behavioral habits of long-tongued hawkmoths, and only to such moths among all the types of flower-visiting animals in their environment. Furthermore, hawkmoths are observably attracted to these flowers and do pollinate them with a high degree of effectiveness. As regards the origin of the flowers of *I. candida* and *I. thurberi*, the simplest interpretation of the observational evidence which is consistent with modern evolutionary theory* is that these flowers have been molded to fit hawkmoths by natural selection for hawkmoth pollination.

Of course, in the long-term span of evolution, hawkmoths have been perfected by natural selection for the ability to feed successfully on the nectar of flowers, just as flowers of a certain type have been developed for hawkmoth pollination. The evolution of flowers and of flower-visiting animals has been a stepwise process involving mutual interactions between the plants and animals at each step along the way, as was clearly recognized by the classical floral biologists of the last century. But we are not concerned in the present discussion with the whole long-term, two-sided process of flower-insect evolution.

Our problem is to explain the latest chapter of flower evolution in some specific cases in the Polemoniaceae. For this purpose it is convenient and adequate

* Modern evolutionary theory is not accepted by some students of flower pollination in Europe, and this is the crucial point in the floral biology controversy, but the point is obviously too far-reaching to deal with here. See Stebbins (1957), Pijl (1958), and Grant (1958) for brief critiques of selected works by Nelson, Good, Werth, and Vogel.

to regard the existing array of flower-visiting animals in a territory as a part of the environment of the plants, to which the latter must respond adaptively if they are to reproduce successfully. Returning to our example, *Celerio lineata* and other similar hawkmoths range widely in North America and elsewhere, and feed on the flowers of many species of plants outside the genus Ipomopsis; whereas *I. candida* and *I. thurberi* are cross-pollinated mainly by these sphingids, and are adapted rather specifically to them.

The relationship between the floral mechanism of *I. candida* and *I. thurberi*, on the one hand, and *Celerio lineata* and other similar hawkmoths, on the other, is like a lock and key in which the lock has been fitted to a preexisting key. It is a somewhat loose-fitting lock and key, to be sure, but this is an advantageous condition. Hawkmoths exhibit considerable individual and interspecific variation in tongue length, feeding habits, and other relevant characteristics. The lock has therefore certainly been exposed to the operation of different keys, and has probably been selected for the ability to work in relation to an array of keys within certain limits. It is a by-product of this looseness of the lock-and-key relation that some non-adapted keys, as exemplified by small solitary bees, happen to be able to turn the lock.

Many species of Gilia, Eriastrum, Navarretia, and other genera have small funnelform flowers which are visited and pollinated by a fairly wide assortment of small and medium-small bees and beeflies. We cannot say that such flowers are either bee flowers or beefly flowers exclusively, nor that they fit any single size class of these insects exclusively. In fact they are facultative bee and beefly flowers. In their evolution they have probably been shaped by several or many species belonging to each insect group. Indeed some features of the floral mechanism, such as the arrangement of the unequal stamens at different levels inside and outside the corolla throat, function to broaden the range of possible pollinating agents by ensuring pollen deposition on insects of different body size and feeding postures.

Yet the spectrum of coadapted pollinators in these cases, though broad, has its limits, for butterflies and beetles visit the flowers occasionally but do not pollinate them effectively, and other classes of nectar-feeding animals do not visit these flowers at all. The lock-and-key relation is more loose-fitting here than between hawkmoth flowers and hawkmoths in that more keys can operate the lock, but the tolerance is still circumscribed within certain limits which exclude many other types of keys.

The Phlox family as a whole exploits many classes of animals in the service of cross-pollination: large bees and small bees; hawkmoths, noctuids, and butterflies; hummingbirds and bats; beeflies, scavenger flies, and beetles; and others. But no species of flower is known in the family which is or could be pollinated effectively by all these agents. Every known species of Polemoniaceae is specialized for pollination by some particular class of pollinators.

Now a class of pollinators may be a narrow one, as in hawkmoth flowers, or a broad one, as in bee and beefly flowers. Furthermore, as in the latter example, the members of a class do not necessarily have to be related phylogenetically. The essential factor is similarity between different species of pollinators with respect to structural and behavioral characteristics relevant to their efficacy in pollinating a given kind of flower, whether this similarity has come about through common phylogenetic descent or convergence.

Many, if not most, species of flower in the family are also visited and pollinated, at least occasionally, by animals for which they are not specially fitted, for example, hawkmoth flowers by small bees. This situation can lead to dangers of interpretation when too few field observations are made. It is easy to see small bees on *Ipomopsis candida* flowers by day, but considerably more difficult to observe hawkmoths on the same flowers at night. Erroneous conclusions could be drawn from casual observations of *Ipomopsis candida*.

In case after case during this investigation, where a plant species has been studied extensively, the flowers have been found to be associated, not only with poorly coadapted animals, but also with some class of pollinator which fits with a nice correspondence of parts into the floral mechanism. This experience has been repeated enough times now to justify the expectation that other cases still unsolved will with further study fall into the same pattern. A lock-and-key relation between flowers and pollinators, which may be loose-fitting but is always exclusive to some extent, is certainly widespread and probably universal in the Phlox family.

PHYLOGENY OF POLLINATION SYSTEMS

The Polemoniaceae encompasses a wide diversity of vegetative and floral forms and a wide range of conditions from relatively primitive to relatively advanced. The evidence of comparative morphology, cytology, ecology, and geographical distribution supports the grouping of these forms into five main circles of affinity (the five tribes), and into subordinate circles of affinity within each tribe (the genera and sections). The same lines of evidence indicate that some species in each circle of affinity have retained a relatively high proportion of primitive characters, and approach in this respect the comparable primitive members of other related circles of affinity, whereas other species possess a larger share of advanced features.

A phylogenetic hypothesis purporting to express these relationships was set forth in *Natural History of the Phlox Family* (Grant, 1959), to which the reader is referred for the details. This hypothesis cannot claim to have complete certainty, but it does claim to be consistent with the available evidence. That evidence, furthermore, has both weaknesses and strengths, which should be borne in mind in estimating the probable reliability of the phylogeny based on it.

A serious weakness in the evidence is the virtual absence of a fossil record. An asset which compensates to a considerable extent for this weakness is the rich representation of living forms, exhibiting different stages of advancement in various characteristics, with connecting links between many of the sections and genera, and relatively narrow gaps between some of the tribes. Now this propitious condition is found especially in the North American tribes (the Bonplandia, Polemonium, and Gilia tribes). It is not so true of the tropical and Andean tribes (the Cobaea and Cantua tribes), which appear to be the isolated remnants of ancient groups. Therefore, our attempt to reconstruct a probable phylogeny from the comparative biology of living species should and does restrict itself mainly to the North American tribes, concerning which we can hypothesize with adequate reliability.

Several recent authors have emphasized the difficulty of constructing a phylogeny without a fossil record. The difficulties have perhaps been over-emphasized and overgeneralized in the recent discussions. The potential contribution of comparative morphology and comparative ecology to phylogeny tends to be underrated by authors who have not worked in these fields. And it seems to be assumed in these discussions that one can generalize about the difficulties of phylogenetic analysis in a way which will apply to all taxonomic groups alike, with all their diverse taxonomic structures. It may well be impossible to reconstruct the phylogeny of a group which has undergone much extinction and has not left a fossil record; but the situation is by no means the same in a living group caught in its heyday of evolution.

We will now put our phylogenetic hypothesis concerning the evolutionarily active part of the Phlox family to use in the following way. We can correlate our knowledge about breeding systems and modes of pollination in the different species of the North American Polemoniaceae with the phylogenetic position of these species. This position, it should be noted, is inferred from evidence independent of the floral ecological findings. From this correlation we can then deduce a probable phylogeny of the pollination systems in this branch of the family.

Cross-pollination by bees is the most widespread condition in the North American Polemoniaceae. It is the one pollination system common to the primitive members of various genera in each of the three North American tribes. Through bee pollination, and through it alone, the divergent tribes and the divergent sectional and generic lines within the tribes can be linked together. Therefore bee pollination is inferred to be the probable common ancestral condition in the North American Polemoniaceae, and all other pollination systems in this group are considered to be derived.

Hummingbird pollination has arisen independently in four separate genera of North American Polemoniaceae in which bee pollination is the basic condition, viz., Loeselia, Gilia, Ipomopsis, and Polemonium. The convergent hummingbird flowers are shown in Plate II.

This indication of repeated transitions from bee to bird pollination is part of a larger pattern in the North American flora. Basically bee pollinated genera in other families in this flora have likewise given rise to hummingbird pollinated species. As examples we may cite Aquilegia (Ranunculaceae), Delphinium (Ranunculaceae), Penstemon (Scrophulariaceae), Mimulus (Scrophulariaceae), Antirrhinum-Galvesia (Scrophulariaceae), Salvia (Labiatae), Monardella (Labiatae), and Brodiaea (Liliaceae).

In the family Polemoniaceae as a whole, hummingbird flowers are found in two additional genera in the Cantua tribe, viz., Cantua and Huthia. As was pointed out earlier, we do not have an adequate representation of living groups in this branch of the family for the employment of the comparative method and hence for drawing phylogenetic inferences. We cannot safely conclude that hummingbird pollination is a derived condition—in fact it could as well have been original—in the tropical and Andean tribes.

Long-tongued fly flowers are closely related to bee flowers and apparently derived from them in the genera Gilia and Linanthus and in the suprageneric lines Polemonium-Collomia-Allophyllum and Eriastrum-Langloisia. Among long-tongued fly flowers, those with very long nectar tubes adapted to Eulonchus, as found in Gilia and Linanthus, clearly represent an extreme development from the related beefly flowers with moderately long tubes.

Scavenger fly pollination in Polemonium, and perhaps also in Ipomopsis, is connected with bee pollination by transitional stages and is evidently derived from it. The same is true of beetle pollination as found in Ipomopsis and Linanthus.

The genera Phlox and Leptodactylon are separated by fairly large gaps from their nearest more primitive relatives in the Polemonium and Gilia tribes respectively. The butterfly flowers in these genera are therefore also isolated from the more primitive pollination systems. Without a series of stages to judge from, we cannot tell whether butterfly pollination in Phlox and Leptodactylon developed directly or indirectly via some intermediate condition from bee pollination.

Noctuid moth pollination in one specialized branch of the genus Phlox can be regarded most probably as an advancement over butterfly pollination in the same genus.

Hawkmoth flowers are found as an advanced condition in Loeselia, Phlox, Ipomopsis, and Linanthus, among the North American tribes, as well as in Cobaea in the tropical Cobaea tribe. In Phlox hawkmoth flowers are connected by transitional stages with less advanced butterfly flowers. In the other genera the situation is less clear. It is possible that moth flowers have developed from bee flowers in Linanthus and Loeselia.

The origin of bat pollination in Cobaea, and for that matter in other plant families, is an interesting but poorly understood problem. Bat pollination is clearly a highly derived condition in the angiosperms (Baker and Harris, 1957). In Cobaea this condition happens to occur in a genus which occupies an isolated position in its family. The absence of connecting links with other more primitive pollination systems makes phylogenetic inferences impossible in this case. It may be worth noting, however, that in several other tropical plant groups both bat and bird pollination have been reported or predicted for different members of the same genus. Some examples are Durio (Bombacaceae), Parkia (Leguminosae), Marcgravia (Marcgraviaceae), Crescentia (Bignoniaceae), and Musa (Musaceae).

Autogamy is found in some 11 genera of the North American Polemoniaceae, where it is always associated with an annual life cycle and reduced floral and vegetative structures. This derived condition has thus arisen independently many times in the history of the temperate Polemoniaceae. It is derived most frequently from bee pollination, as in Gilia and Polemonium. Long-tongued fly flowers have also given rise to autogamous offshoots in Langloisia and Linanthus. The autogamous minor genus Microsteris is a reduced derivative of Phlox, which is Lepidoptera-pollinated.

The systematic distribution of self-incompatibility is interesting. Self-incompatibility has a spotty occurrence in the family, whereas self-compatibility is widespread and common. In some groups, moreover, self-incompatibility

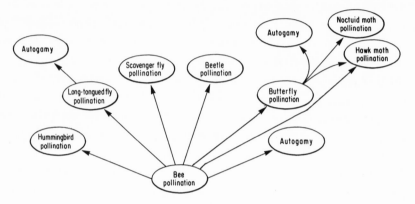

Fig. 46. Probable phylogeny of pollination systems in the main North American branch of the Phlox family.

occurs sporadically in the more advanced members, while the more primitive members have an outcrossing floral mechanism along with a self-compatible breeding system. In such cases self-incompatibility is clearly a derived condition. Since the foregoing trend does not hold true in all groups, however, but is subject to exceptions, it is not possible to decide whether self-incompatibility or self-compatibility with predominant outcrossing is the original condition in the North American branch of the family as a whole.

The conclusions reached in the foregoing discussion are summarized diagrammatically in Fig. 46, which shows the suggested phylogenetic relations between the various pollination systems in the North American tribes of the Phlox family.

LEVELS OF DIFFERENTIATION

The adaptive radiation in mode of pollination did not take place once and for all at some one period in the past history of the Polemoniaceae. It has been a continuing process. We infer this from the following considerations.

If the adaptive radiation had been initiated and completed long ago, we would expect to find one whole major group composed of, say, bee flowers and another related major group composed uniformly of, say, Lepidoptera flowers. As a matter of fact, we do find this situation in some cases. Thus bee pollination predominates in the genus Polemonium and Lepidoptera pollination in the genus Phlox. But we also find the differentiation at lower taxonomic levels in other parts of the family.

In Linanthus the divergence in respect to pollination appears at the section level. Section Pacificus contains bumblebee flowers; section Dactylophyllum has small bee flowers; section Leptosiphon long-tongued fly flowers; and section Linanthus moth flowers.

In Ipomopsis, species and species groups differ in mode of pollination. Bee flowers are found in *I. multiflora*; hummingbird flowers in *I. aggregata* and *I. rubra*; butterfly flowers in *I. macombii*; hawkmoth flowers in *I. candida* and *I. tenuituba*; scavenger fly flowers in *I. spicata*; beetle flowers in *I. congesta*; and autogamous flowers in *I. pumila* and its relatives.

And in Gilia and some other genera we find races of the same species which differ in method of pollination. The case of *Gilia splendens* is interesting in this connection. This annual Gilia with funnelform pink flowers occurs in openings in pine woods from central to southern California. The widespread race with medium-long nectar tubes is pollinated mainly by Bombylius; the high San Gabriel Mt. race with long slender tubes by Eulonchus; a localized race in the San Bernardino Mts. with long stout tubes by hummingbirds; and a desert race with smaller dull-colored flowers by autogamy (see Fig. 15). These races are known to replace one another geographically, to intergrade along some transects, and to be interfertile in the experimental garden (V. and A. Grant, 1954).

Such cases of pollination races are not uncommon. They have turned up in species after species in Gilia, Eriastrum, Allophyllum, and other genera. The discovery that geographical variation in some species of Polemoniaceae has a component of adaptation to different kinds of pollinators indicates that the plant populations are responsive to the array of flower-visiting animals in their respective territories.

We even begin to find indications of parallel racial variation in separate species with respect to pollination. In coastal California both *Eriastrum densifolium* and *Gilia achilleaefolia* are represented by races with broad-throated flowers in maritime districts and by small-throated races in the dry interior mountains. In each of these species the maritime race is pollinated by large bees and the interior race by small bees and beeflies.

Where an adaptive radiation is in its initial stages we can study it as a process rather than as a historical event. Particular interest therefore attaches to cases of racial and specific differentiation in method of pollination.

THE PROCESS OF DIFFERENTIATION

Many populations in the Polemoniaceae are pollinated by agents belonging to two or more taxonomic groups and differing in their structural and behavioral characteristics. The combination of bee and beefly pollination is found within a given population in many species of Gilia. Other combinations are bee and scavenger fly pollination (*Polemonium delicatum*); bee and hummingbird pollination (*Polemonium confertum*); and hummingbird and hawkmoth pollination (*Cantua quercifolia*). A mixture of insect cross-pollination and autogamy is common in many annual species of the North American tribes.

Adaptation to a wide range of pollinators has both advantages and disadvantages. In a plant population which is pollinated by a variety of agents, seed production is independent of seasonal fluctuations in the visitations by any single type of agent, and thus tends to be stable and reliable from year to year. On the other hand, on the principle that the jack-of-all-trades is master of none, a floral mechanism which is broadly adapted to many classes of visitors with different structural and behavioral characteristics may not be pollinated with high effectiveness by any single agent. In practice a compromise is no doubt reached between the extremes of broad adaptation and narrow specialization. A population of flowers probably becomes adapted to as broad a range of pollinating agents in its territory as it can attract and utilize with effectiveness in pollination.

Now the spectrum of flower-visiting animals undergoes changes along an ecogeographical transect. Scavenger flies increase in importance relative to bees as flower pollinators at high elevations in the western American mountains as well as in the far north. In a similar manner the importance of bumblebees relative to small solitary bees increases from the dry interior toward the moist coastline in central and southern California. The climate of pollinators, as we may term it, changes from one area to another.

As a result, two populations of the same plant species living in different territories may both receive effective pollinating visits from the same two classes of pollinators, say bees and beeflies, but receive the two types of visitations in different relative frequencies. If one local race receives a greater number of effective bee visits and the other population a greater number of effective beefly visits, natural selection will favor closer floral adaptations to the special character-istics of bees in the first race and closer coadaptations with beeflies in the second. The point of compromise in floral adaptation will shift in correspondence with the climate of pollinators in each given territory.

The different races of *Gilia leptantha* are visited and pollinated by both bees and beeflies. In *G. l. pinetorum* the broad funnelform corolla favors feeding and pollination by the bees, without excluding the beeflies. The subsalverform flowers of *G. l. leptantha* and *G. l. purpusii* with their long slender nectar tube are fitted more especially for beeflies and perhaps for other long-tongued flies, but not to the point of excluding pollen-collecting bees. The racial differences in mode of pollination are relative.

A normally outcrossing population of *Gilia achilleaefolia achilleaefolia* is self-compatible but suffers from inbreeding depression. Experimental lines derived from this natural population were self-pollinated for several generations and their inbred progeny in each generation were selected for vigor. Homozygous segregants were obtained after several generations of artificial selection which no longer showed inbreeding depression. In this respect the derived lines were like the natural populations of another race of the same species, *G. a. multicaulis*, which is predominantly autogamous and can maintain normal vigor under continuous inbreeding.

One population of *Gilia leptantha leptantha* is known to reproduce partly by insect pollination and partly by autogamous self-pollination. This polymorphic population contains both autogamous and non-autogamous individuals. Seeds collected at random in the natural population served as the starting point of an experimental population which was grown in an insect-free screenhouse for several generations and allowed to set seeds by its own devices. Under the experimental conditions the non-autogamous types were quickly lost and the derivative population came to consist wholly of autogamous individuals.

Similar changes have taken place naturally in the races of numerous species which have colonized desert or high mountain or other habitats where insect visits are scarce or unreliable. The transition to autogamy has been particularly common in annual species where, as is well known, each individual has but a single flowering season in which to produce seeds, and often a quite brief flowering season at that. The autogamous method of seed formation comes to have an extremely high selective value in annual plants under conditions of unreliable insect activity. This principle, predicted by Stebbins (1950, 176) on theoretical grounds, is abundantly confirmed by the empirical evidence in the Phlox family.

Speciation being an extension of race formation, we would expect to find cases in which the differences between related species in mode of pollination are more exclusive than are those between races. Such cases are found commonly.

The horizontal, broad-throated, bilaterally symmetrical flowers of *Allophyllum glutinosum* provide a good perching place for the bees which feed on and pollinate them. The closely related *Allophyllum divaricatum* has slender salverform flowers which are fitted primarily for long-tongued flies and are not well suited for bee visits. *Linanthus grandiflorus* with its full-throated flowers borne in heads is a bumblebee flower, while *Linanthus androsaceus* and *L. parviflorus* with long, slender, salverform flowers are specialized for long-tongued flies in ways which virtually exclude visits by heavy bees.

When the specializations for different classes of pollinators approach or reach a stage of mutual exclusiveness, these differences contribute to the reproductive isolation between the species involved, and permit such species to evolve henceforth along genetically independent lines. The bumblebee-pollinated *Linanthus grandiflorus* and the long-tongued-fly–pollinated *Linanthus androsaceus*, which occur in neighboring habitats on the California coast, are thus isolated mechanically (and no doubt by other isolating mechanisms too). Floral isolation exists between the related species *Ipomopsis aggregata* and *I. tenuituba* in the Rocky Mt. region, which are normally pollinated by hummingbirds and by hawkmoths respectively. Autogamy may be differentiated from insect pollination not only at the race level, as discussed earlier, but also at the species level, as exemplified by the autogamous *Gilia millefoliata* and the related bee-pollinated *Gilia capitata* on the coastal strand of northern California.

We find a certain diversity in pollination systems in any area which is inhabited by numerous species of Polemoniaceae. A typical situation in some regions of California, for example, is the occurrence of several polemoniaceous species of bee flowers and facultative bee and beefly flowers, along with about one species each of long-tongued fly flower, moth flower, and hummingbird flower, and several autogamous species. The process of specialization for different classes of pollinators within an area, which enables the sympatric species to fit into different facies of the same climate of pollinators, can be explained as a result of interspecific competition for pollinating agents.

The bee fauna in an area can probably provide for the pollination of many, but not an unlimited number of, plant species. The polemoniaceous species possessing the supposed ancestral system of bee pollination do coexist in fair numbers in many areas in western North America. These Polemoniaceae and the unrelated species of bee flowers in other plant families in the same community must often utilize the existing supply of bees to full capacity. The occasional finding of two closely sympatric populations of bee flowers, only one of which is actively visited by bees and sets abundant seeds, indicates that the postulated situation does indeed arise. Such conditions of interspecific competition for bees would favor the emergence of specializations for other types of pollinators, as yet relatively little exploited, even though they might be intrinsically less efficient as pollen carriers than the bees.

Autogamy can be viewed, not only as an adaptation to a sparse climate of pollinators, but also and more generally as a successful means of escape from the competition for pollinating insects which is of course especially critical in a region where such insects are scarce. This may be the reason why the self-pollinating species, which do not enter into a direct competition for animal visitors, can and often do coexist in larger numbers in a territory than the species possessing any single method of cross-pollination.

It appears likely in theory, as we have stated earlier, that a population of flowers is exposed to opposing selective pressures for broad adaptation versus narrow specialization in relation to pollination. In each particular case, moreover, these opposing pressures probably approach an equilibrium point in which the net advantages of broad adaptation and of narrow specialization prevail over their respective disadvantages. The life of flowers is a practical compromise between opposite extremes. Now we can suggest further that an important factor—perhaps the chief factor—affecting the point of equilibrium is interspecific competition for pollinating agents. As the interpecific competition for pollinators increases in strength, the compromise point can be expected to shift toward more narrow specialization in the floral mechanism and pollination system.

In summary, the diversity of flower forms in the Phlox family is the result of an adaptive radiation in the mode of pollination, and this adaptive radiation in turn is an outcome of the processes involved in speciation.

SYSTEMATIC LIST OF ANIMAL VISITORS

I. Bees (*Apoidea*)

Fig. 30

Apidae-Apinae

Apis mellifera (introduced in area of following plant species): Polemonium reptans, Polemonium caeruleum, Gilia angelensis, Gilia achilleaefolia, Gilia capitata, Gilia latiflora, Gilia leptantha, Gilia splendens, Eriastrum densifolium, Eriastrum sapphirinum

Bombus alticola: Polemonium caeruleum

Bombus americanorum: Polemonium reptans, Phlox divaricata, Phlox pilosa

Bombus auricomus: Polemonium reptans, Phlox pilosa

Bombus balteatus: Polemonium boreale

Bombus centralis: Gilia pinnatifida

Bombus consimilis: Phlox divaricata

Bombus edwardsii: Gilia capitata

Bombus flavifrons: Polemonium foliosissimum, Gilia pinnatifida

Bombus griseocollis: Polemonium reptans, Phlox pilosa

Bombus hortorum: Polemonium caeruleum

Bombus impatiens: Polemonium reptans, Phlox divaricata, Phlox pilosa

Bombus lapidarius: Polemonium caeruleum

Bombus lapponicus: Polemonium caeruleum

Bombus lucorum: Polemonium caeruleum

Bombus melanopygus: Gilia pinnatifida

Bombus muscorum: Cobaea scandens

Bombus pratorum: Polemonium caeruleum

Bombus ridingsii: Polemonium reptans, Phlox divaricata

Bombus rufocinctus: Polemonium foliosissimum, Polemonium pulcherrimum, Polemonium confertum

Bombus separatus: Phlox divaricata

Bombus sonorus: Ipomopsis multiflora, Ipomopsis thurberi

Bombus terrestris: Polemonium caeruleum

Bombus vagans: Polemonium reptans, Phlox divaricata

Bombus virginicus: Phlox divaricata

Bombus vosnesenskii: Gilia capitata, Eriastrum densifolium, Linanthus grandiflorus

Bombus sp.: Polemonium caeruleum, Polemonium boreale, Polemonium pulcherrimum, Polemonium viscosum, Polemonium eximium, Ipomopsis aggregata.

Psithyrus variabilis: Polemonium reptans, Phlox divaricata

Apidae-Xylocopinae

Ceratina calcarata: Polemonium reptans

Ceratina dupla: Polemonium reptans

Ceratina sp.: Allophyllum glutinosum, Gilia capitata, Linanthus liniflorus

Apidae-Anthophorinae

Anthophora flexipes: Eriastrum sapphirinum

Anthophora urbana: Gilia diegensis, Gilia leptantha, Eriastrum densifolium

Anthophora ursina: Polemonium reptans, Phlox pilosa

Apidae-Anthophorinae (*Cont.*)

Anthophora sp.: Allophyllum glutinosum, Gilia latiflora, Eriastrum densifolium, Eriastrum sapphirinum, Langloisia matthewsii, Navarretia squarrosa

Diadasia sp.: Eriastrum sapphirinum

Emphoropsis sp.: Gilia latiflora

Exomalopsis sp.: Navarretia hamata

Nomada affabilis: Polemonium reptans

Nomada hydrophylli: Polemonium reptans

Nomada ovata: Polemonium reptans

Nomada superba: Phlox pilosa

Nomada sp.: Gilia capitata

Oreopasites sp.: Navarretia atractyloides, Linanthus liniflorus

Tetralonia belfragei: Polemonium reptans

Tetralonia californica: Gilia latiflora

Tetralonia dilecta: Polemonium reptans, Phlox divaricata, Phlox pilosa

Tetralonia sp.: Gilia tricolor, Gilia latiflora

Megachilidae

Anthidium flavipes: Phlox paniculata

Anthidium maculosum: Ipomopsis macombii

Anthidium strigatum: Phlox paniculata

Ashmeadiella bigeloviae: Eriastrum sapphirinum

Ashmeadiella bucconis denticulata: Navarretia viscidula

Ashmeadiella californica: Gilia angelensis

Ashmeadiella californica californica: Eriastrum sapphirinum, Navarretia viscidula

Ashmeadiella gillettei cismontanica: Eriastrum sapphirinum

Ashmeadiella meliloti meliloti: Eriastrum sapphirinum

Ashmeadiella sonora: Eriastrum sapphirinum

Ashmeadiella sp.: Eriastrum sapphirinum, Eriastrum luteum, Navarretia atractyloides, Linanthus liniflorus

Calanthidium sp.: Linanthus liniflorus

Chelostoma campanularum: Polemonium caeruleum

Chelostoma cockerelli: Gilia exilis

Chelostoma incisulum: Gilia tricolor

Chelostoma nigricorne: Polemonium caeruleum

Chelostoma sp.: Gilia angelensis, Linanthus liniflorus

Chelostomopsis rubifloris: Gilia tricolor, Gilia exilis, Gilia diegensis

Coelioxys sp.: Polemonium caeruleum

Dianthidium ulkei: Gilia leptalea

Dianthidium sp.: Ipomopsis macombii

Hoplitis colei: Eriastrum sapphirinum

Hoplitis fulgida: Polemonium foliosissimum, Polemonium delicatum

Hoplitis grinnellii: Eriastrum sapphirinum

Hoplitis producta: Gilia exilis

Megachile melanophaea: Gilia pinnatifida

Megachile sp.: Polemonium caeruleum, Gilia latiflora

Osmia atriventris: Polemonium reptans

Osmia conjuncta: Polemonium reptans

Osmia cordata: Phlox divaricata

Osmia distincta: Gilia tricolor

Osmia lignaria: Polemonium reptans

Osmia pumila: Polemonium reptans

Osmia rufa: Polemonium caeruleum

Osmia sp.: Polemonium pulcherrimum, Polemonium confertum, Gilia capitata, Gilia ochroleuca, Ipomopsis congesta, Navarretia atractyloides

Robertsonella simplex: Polemonium reptans

Melittidae

Hesperapsis sp.: Gilia ochroleuca

Halictidae

Agapostemon cockerelli: Gilia leptantha

Agapostemon radiatus: Polemonium reptans

Agapostemon texanus: Phlox nana

Agapostemon sp.: Gilia capitata

Augochlora pomoniella: Gilia achilleaefolia

Augochlora pura: Polemonium reptans

Augochlora similis: Polemonium reptans

Augochlora striata: Polemonium reptans

Augochlora sp.: Ipomopsis spicata

Chloralictus sp.: Allophyllum divaricatum, Gilia tricolor, Gilia angelensis, Gilia leptantha, Ipomopsis aggregata, Ipomopsis tenuituba, Ipomopsis candida, Ipomopsis congesta, Ipomopsis spicata

Dialictus sp.: Eriastrum luteum

Halictidae   (*Cont.*)

Dufourea linanthi: Linanthus liniflorus

Dufourea versatilis: Gilia leptalea

Dufourea sp.: Gilia tricolor, Gilia angelensis, Gilia achilleaefolia, Gilia capitata, Gilia cana, Gilia diegensis (2 spp.), Gilia brecciarum, Gilia malior, Gilia splendens, Linanthus liniflorus, Linanthus dianthiflorus

Evylaeus sp.: Gilia leptantha

Halictus farinosus: Gilia achilleaefolia, Gilia splendens, Eriastrum densifolium

Halictus rubicundus: Polemonium reptans

Halictus smeathmanellus: Phlox paniculata

Halictus tripartitus: Gilia caruifolia

Halictus subgenus Halictus: Gilia tricolor, Gilia angelensis, Gilia capitata

Halictus subgenus Seladonia: Gilia angelensis, Gilia capitata, Linanthus liniflorus

Halictus sp.: Gilia ochroleuca

Lasioglossum coreopsis: Polemonium reptans

Lasioglossum coriaceum: Polemonium reptans

Lasioglossum kincaidii: Gilia capitata

Lasioglossum obscurum: Polemonium reptans

Lasioglossum pilosum: Polemonium reptans

Lasioglossum sisymbrii: Gilia capitata, Gilia splendens, Eriastrum densifolium

Lasioglossum trizonatum: Gilia leptantha

Lasioglossum versatum: Polemonium reptans

Unidentified: Gilia tricolor

Andrenidae

Ancylandrena sp.: Navarretia atractyloides

Andrena atala: Polemonium foliosissimum

Andrena carlini: Polemonium reptans

Andrena chapmanae: Gilia cana

Andrena geranii: Polemonium reptans

Andrena nasonii: Polemonium reptans

Andrena polemonii: Polemonium reptans

Andrena pruni: Polemonium reptans

Andrena prunorum: Gilia capitata

Andrena sayi: Polemonium reptans

Andrena sp.: Gilia tricolor, Gilia achilleaefolia, Gilia capitata, Gilia diegensis, Navarretia pubescens, Linanthus parviflorus

Nomadopsis barbata: Linanthus dianthiflorus

Nomadopsis sp.: Gilia capitata

Perdita giliae: Ipomopsis aggregata

Perdita pelargoides: Eriastrum sapphirinum, Navarretia atractyloides

Perdita tristella: Eriastrum sapphirinum

Perdita sp.: Ipomopsis macombii, Eriastrum sapphirinum, Linanthus dianthiflorus

Unidentified: Gilia tricolor

Colletidae

Hyalaeus sp.: Gilia achilleaefolia, Ipomopsis congesta

Unidentified bees: Polemonium californicum, Polemonium pulcherrimum, Microsteris gracilis, Gilia splendens, Eriastrum densifolium, Cantua quercifolia

II. Wasps (Other Hymenoptera)

Braconidae

Unidentified: Ipomopsis congesta, Linanthus liniflorus

Sphecidae

Ammophila sp.: Ipomopsis congesta

Scoliidae

Campsomeris ephippium: Cantua quercifolia

Vespidae

Polybia occidentales: Cantua quercifolia

Vespa germanica: Cobaea scandens

III. Butterflies and Skippers (Rhopalocera)   Figs. 7, 18, 34

Papilionidae

Papilio· ajax: Phlox glaberrima, Phlox divaricata, Phlox pilosa

Papilio cresphontes: Phlox glaberrima, Phlox divaricata

Papilio glaucus: Phlox divaricata, Phlox pilosa

Papilionidae (*Cont.*)
  Papilio marcellus: Phlox divaricata
  Papilio philenor: Phlox glaberrima, Phlox divaricata, Phlox drummondii, Ipomopsis macombii
  Papilio rutulus: Phlox paniculata, Gilia achilleaefolia, Eriastrum densifolium, Leptodactylon californicum
  Papilio thoas: Phlox glaberrima, Phlox divaricata
  Papilio troilus: Phlox divaricata, Phlox pilosa

Pieridae
  Colias eurytheme: Phlox pilosa, Gilia tricolor, Gilia capitata
  Colias philodice: Polemonium reptans, Phlox glaberrima, Phlox divaricata, Phlox pilosa
  Eurema sp.: Eriastrum densifolium
  Pieris rapae: Gilia capitata, Eriastrum densifolium
  Unidentified: Gilia achilleaefolia

Danaidae
  Danaus plexippus: Phlox glaberrima, Phlox divaricata, Phlox multiflora

Satyridae
  Coenonympha californica: Gilia capitata

Nymphalidae
  Argynnis semiramis: Eriastrum densifolium
  Dione vanillae comstocki: Gilia achilleaefolia
  Euphydryas chalcedona: Gilia capitata, Navarretia hamata
  Melitaea acastus: Ipomopsis congesta
  Melitaea gabbii: Gilia capitata
  Phyciodes myllita: Gilia capitata
  Phyciodes tharos: Phlox pilosa
  Speyeria cybele: Phlox pilosa
  Vanessa cardui: Eriastrum densifolium
  Vanessa carye: Gilia angelensis, Gilia achilleaefolia, Gilia cana

Vanessa virginiensis: Phlox pilosa
Vanessa sp.: Phlox pilosa, Linanthus grandiflorus

Lycaenidae
  Lycaena gorgon: Eriastrum densifolium
  Lycaena thoe: Phlox pilosa
  Plebius icarioides: Gilia capitata
  Unidentified: Ipomopsis congesta, Eriastrum sapphirinum, Leptodactylon californicum

Hesperiidae
  Epargyreus tityrus: Phlox divaricata, Phlox pilosa
  Erynnis brizo: Polemonium reptans
  Erynnis callidus: Linanthus nuttallii
  Erynnis icelus: Phlox divaricata
  Erynnis juvenalis: Polemonium reptans
  Erynnis propertius: Linanthus nuttallii
  Hesperia harpulus leussleri: Eriastrum densifolium
  Hesperia juba: Gilia capitata
  Hylephila phylaeus: Gilia angelensis
  Ochlodes snowi: Ipomopsis macombii
  Ochlodes sylvanoides: Leptodactylon californicum
  Poanes hobomok: Phlox divaricata
  Poanes zabulon: Phlox divaricata
  Polites peckius: Phlox glaberrima, Phlox pilosa
  Polites themistocles: Phlox divaricata
  Thorybes bathyllus: Phlox divaricata, Phlox pilosa
  Thorybes pylades: Phlox pilosa
  Unidentified: Phlox diffusa, Gilia capitata, Leptodactylon californicum

Unidentified butterflies: Phlox paniculata, Phlox maculata, Phlox stolonifera, Phlox subulata, Phlox pilosa, Phlox roemeriana, Phlox andicola, Eriastrum sapphirinum

IV. *Moths* (*Heteroneura*)   Figs. 9, 17, 24

Sphingidae
  Celerio lineata: Phlox divaricata, Phlox

drummondii, Gilia latiflora, Ipomopsis aggregata, Ipomopsis tenuituba, Ipomopsis candida,

Sphingidae   (*Cont.*)

Ipomopsis thurberi, Ipomopsis longiflora, Langloisia punctata, Leptodactylon californicum, Linanthus dichotomous

Cocytius sp.: Cobaea penduliflora

Dolba hylaeus: Phlox paniculata

Hemaris gracilis: Phlox paniculata

Hemaris senta: Phlox paniculata, Leptodactylon californicum

Hemaris thysbe: Polemonium reptans, Phlox divaricata

Macroglossa stellatarum: Phlox paniculata, Cobaea scandens

Phlegethontius sp.: Cobaea penduliflora

Proserpinus clarkiae: Leptodactylon californicum

Sphinx drupiferarum: Ipomopsis candida

Xylophanes sp.: Cobaea penduliflora

Unidentified: Cantua quercifolia

Noctuidae

Euxoa messoria: Phlox caespitosa

Heliothis obsoleta: Phlox diffusa

Plusia gamma: Phlox paniculata

Syntomidae

Scepsis fulvicollis: Phlox glaberrima

Phalaenidae

Autographa falcifera simplex: Polemonium reptans, Phlox divaricata, Phlox pilosa

Geometridae

Neoterpes trianguliferata costinotata: Linanthus nuttallii

Pyralidae

Crambus pascuellus: Linanthus nuttallii

Ephestiodes gilvescentella: Linanthus nuttallii

Adelidae

Adela simpliciella; Gilia achilleaefolia, Linanthus parviflorus

Adela sp.: Gilia caruifolia

Unidentified moths: Ipomopsis aggregata

V. *Long-Tongued Flies*   *Fig.* 31

Bombyliidae

Anastoechus barbatus: Navarretia peninsularis

Anastoechus sp.: Linanthus liniflorus

Aphoebantus sp.: Allophyllum glutinosum, Gilia latiflora, Gilia splendens

Bombylius atriceps: Phlox divaricata, Phlox pilosa

Bombylius lancifer: Allophyllum divaricatum, Gilia tricolor, Gilia achilleaefolia, Gilia capitata, Gilia ochroleuca, Gilia cana, Gilia brecciarum, Gilia splendens, Gilia capillaris, Ipomopsis spicata, Eriastrum densifolium, Navarretia hamata, Navarretia peninsularis, Leptodactylon californicum, Linanthus nuttallii, Linanthus parviflorus, Linanthus breviculus

Bombylius major: Polemonium reptans, Gilia tricolor

Bombylius sp.: Allophyllum violaceum, Gilia achilleaefolia, Gilia cana, Gilia latiflora, Gilia leptantha, Linanthus bicolor

Conophorus sp.: Gilia tricolor

Eclimus luctifer: Linanthus nuttallii

Lepidanthrax inaurata: Gilia leptalea

Lepidanthrax sp.: Allophyllum divaricatum, Eriastrum eremicum, Navarretia atractyloides, Navarretia hamata, Navarretia peninsularis

Lordotus albidus: Gilia latiflora

Lordotus planus: Navarretia hamata

Oligodranes cinctura: Gilia cana, Gilia latiflora

Oligodranes setosus: Gilia leptalea

Oligodranes sp.: Gilia cana, Gilia latiflora, Gilia tenuiflora, Gilia leptantha, Gilia sinuata, Ipomopsis aggregata, Eriastrum luteum

Pantarbes pasio: Eriastrum eremicum

Pantarbes sp.: Gilia tricolor, Ipomopsis congesta, Langloisia matthewsii

Phthiria sp.: Gilia cana, Eriastrum luteum, Linanthus liniflorus

Villa agrippina: Linanthus parviflorus

Bombyliidae (*Cont.*)
Villa alternata: Gilia leptantha, Eriastrum densifolium, Linanthus breviculus
Villa fulviana: Eriastrum densifolium
Villa morio: Eriastrum densifolium
Villa sinuosa jaenickiana: Collomia mazama, Ipomopsis macombii
Villa sp.: Gilia cana

Unidentified: Phlox subulata, Langloisia punctata

Cyrtidae
Eulonchus smaragdinus: Gilia splendens, Leptodactylon californicum, Linanthus androsaceus

VI. *Other Flies (General Diptera)* Fig. 32

Empididae
Empis sp.: Polemonium californicum
Pachymeria pudica: Polemonium reptans
Unidentified: Polemonium delicatum

Syrphidae
Baccha sp.: Ipomopsis aggregata
Eristalis tenax: Phlox paniculata, Gilia angelensis, Cobaea scandens
Eristalis sp.: Polemonium caeruleum
Mesogramma marginata: Polemonium reptans
Platycheirus manicatus: Polemonium caeruleum
Pipiza femoralis: Polemonium reptans
Rhingia campestris: Polemonium caeruleum
Rhingia nasica: Polemonium reptans
Syrphus americanus: Phlox glaberrima, Gilia angelensis, Gilia capitata, Gilia splendens
Syrphus ribesii: Polemonium caeruleum
Syrphus sp.: Polemonium caeruleum, Polemonium pulcherrimum
Volucella bombylans: Polemonium caeruleum
Unidentified: Polemonium delicatum, Polemonium californicum, Polemonium viscosum, Polemonium eximium, Polemonium elegans, Gilia angelensis, Gilia splendens, Ipomopsis spicata, Eriastrum densifolium, Navarretia hamata, Linanthus nuttallii

Conopidae
Conops flavipes: Phlox paniculata
Unidentified: Gilia tricolor, Linanthus liniflorus

Calliphoridae
Protophormia terraenovae: Polemonium viscosum
Unidentified: Polemonium boreale

Anthomyiidae
Unidentified: Polemonium delicatum

Muscidae
Pogonomyia spinitarsus: Polemonium viscosum
Thricops septentrionalis: Polemonium delicatum
Thricops aff. villicrurus: Polemonium viscosum

Tachinidae
Athanatus californicus: Ipomopsis spicata
Chaetogaedia monticola: Linanthus nuttallii
Echinomyia fera: Phlox paniculata
Peleteria sp.: Gilia capitata
Unidentified: Ipomopsis spicata

Unidentified flies: Polemonium boreale

VII. *Beetles (Coleoptera)* Fig. 33

Telephoridae (Cantharidae)
Dasytes flavipes: Polemonium caeruleum

Melyridae
Eschatocrepis constrictus: Gilia achilleaefolia, Ipomopsis congesta

Listrus famelicus: Gilia caruifolia
Malachius mirandus: Gilia tricolor
Trichochrous suturalis: Gilia capitata, Gilia leptantha
Trichochrous sp.: Ipomopsis congesta, Linanthus dianthiflorus, Linanthus parryae

Unidentified: Gilia angelensis, Gilia achilleaefolia, Gilia capitata, Gilia cana, Gilia
brecciarum, Gilia tenuiflora, Gilia capillaris,
Ipomopsis congesta (2 spp.), Langloisia
punctata, Linanthus nuttallii, Linanthus
parviflorus, Linanthus liniflorus (2 spp.)

Cleridae
Trichodes ornatus: Gilia capitata

Buprestidae
Anthaxia aenogaster: Gilia capitata, Gilia
splendens, Linanthus nuttallii

Coccinellidae
Megilla maculata: Polemonium reptans

Mordellidae
Mordella albosuturalis: Gilia capitata, Erias-
trum densifolium
Mordella sp.: Gilia achilleaefolia, Gilia capitata

Meloidae
Macrobasis unicolor: Polemonium foliosissimum

Anthicidae
Corphyra terminalis: Polemonium reptans

Scarabaeidae
Hoplia dispar: Gilia capitata
Serica anthracina: Linanthus grandiflorus

Cerambycidae
Leptura lineola: Gilia capitata
Pachyta interrogationis: Polemonium caeru-
leum

Bruchidae
Acanthoscelides lobatus: Ipomopsis congesta

Unidentified beetles: Polemonium caeruleum

VIII. *Hummingbirds*   *Figs.* 35, 36

Trochilidae
Archilochus colubris: Ipomopsis rubra
Calypte anna: Eriastrum densifolium, Cobaea
scandens
Patagona gigas: Cantua candelilla
Selasphorus platycercus: Ipomopsis aggre-
gata

Selasphorus rufus: Ipomopsis aggregata
Stellula calliope: Polemonium confertum,
Ipomopsis aggregata
Unidentified: Gilia splendens, Ipomopsis
tenuifolia, Ipomopsis arizonica, Ipomopsis
aggregata, Loeselia mexicana, Cantua quercifolia

IX. *Bats*   *Fig.* 37

Phyllostomidae-Glossophaginae
Unidentified: Cobaea trianaei

BIBLIOGRAPHY

ABRAMS, L. 1951. *Illustrated Flora of the Pacific States.* Vol. iii. Stanford Univ. Press, Stanford, Calif.

ALLEN, G. M. 1939. *Bats.* Harvard Univ. Press, Cambridge, Mass.

BAKER, H. G. 1961. The adaptation of flowering plants to nocturnal and crepuscular pollinators. *Quart. Rev. Biol.* **36:** 64–73.

—— 1963. Evolutionary mechanisms in pollination biology. *Science* **139:** 877–883.

BAKER, H. G., and B. J. HARRIS. 1957. The pollination of Parkia by bats and its attendant evolutionary problems. *Evolution* **11:** 449–460.

BEHRENS, W. J. 1880. Der Bestäubungsmechanismus bei der Gattung Cobaea Cavanilles. *Flora* **63:** 403–410.

BRAND, A. 1905. Kulturversuche mit verschiedenen Polemoniaceen-Arten. *Engler's Bot. Jahrb.* **36:** 69–77.

—— 1907. Polemoniaceae. In A. Engler, *Das Pflanzenreich,* **4** (250). Engelmann Verlag, Leipzig.

COCKERELL, T. D. A. 1902. Flowers and insects in New Mexico. *Amer. Nat.* **36:** 809–817.

DARWIN, C. 1862. *On the Various Contrivances by which Orchids are Fertilized by Insects.* John Murray, London.

—— 1877. *The Different Forms of Flowers on Plants of the Same Species.* John Murray, London.

—— 1878. *The Effects of Cross and Self Fertilisation in the Vegetable Kingdom.* 2d ed., John Murray, London.

DAVIDSON, J. F. 1950. The genus Polemonium (Tournefort) L. *Univ. Calif. Publ. Bot.* **23:** 209–282.

EASTWOOD, A. 1893. Field notes at San Emidio. *Zoe* **4:** 144–147.

EKSTAM, O. 1894. Zur Kenntnis der Blütenbestäubung auf Nowaja Semlja. *Öfversigt kongl. Vetenskaps-Akademiens Förhandlingar (Stockholm)* **1894:** 79–84.

—— 1897. Einige blütenbiologische Beobachtungen auf Nowaja Semlja. *Tromsø Museums Aarshefter* **18:** 109–198. [Abstract in *Just's Bot. Jahresber.* **25:** (1): 11–15, 1900.]

—— 1898. Einige blüthenbiologische Beobachtungen auf Spitzbergen. *Tromsø Museums Aarshefter* **20:** 1–66. [Abstract in *Just's Bot. Jahresber.* **26:** (2): 397–401, 1901.]

EPLING, C., H. LEWIS, and F. M. BALL. 1960. The breeding group and seed storage: a study in population dynamics. *Evolution* **14:** 238–255.

ERBE, L., and B. L. TURNER. 1962. A biosystematic study of the *Phlox cuspidata–Phlox drummondii* complex. *Amer. Midl. Nat.* **67:** 257–281.

ERNST, A. 1880. Die Befruchtung von *Cobaea penduliflora*. *Kosmos* **7**: 44–46. (Translation in *Nature* **22**: 148–149, 1880.)

FERGUSON, E. E. 1921. Field notes of the 1920 outing. *Sierra Club Bull.* **11**: 147–150.

FOSBERG, F. R. 1942. Notes on North American plants. III. *Amer. Midl. Nat.* **27**: 761–765.

FRYXELL, P. A. 1957. Mode of reproduction of higher plants. *Bot. Rev.* **23**: 135–233.

GRANT, A., and V. GRANT. 1955. The genus Allophyllum (Polemoniaceae). *Aliso* **3**: 93–110.

GRANT, V. 1950. Genetic and taxonomic studies in Gilia. I. *Gilia capitata*. *Aliso* **2**: 239–316.

—— 1952a. Genetic and taxonomic studies in Gilia. II. *Gilia capitata abrotanifolia*. *Aliso* **2**: 361–373.

—— 1952b. Genetic and taxonomic studies in Gilia. III. The *Gilia tricolor* complex. *Aliso* **2**: 375–388.

—— 1954a. Genetic and taxonomic studies in Gilia. IV. *Gilia achilleaefolia*. *Aliso* **3**: 1–18.

—— 1954b. Genetic and taxonomic studies in Gilia. V. *Gilia clivorum*. *Aliso* **3**: 19–34.

—— 1956. A synopsis of Ipomopsis. *Aliso* **3**: 351–362.

—— 1958. Floral ecology (review). *Ecology* **39**: 778–779.

—— 1959. *Natural History of the Phlox Family*. Systematic Botany. M. Nijhoff, The Hague.

—— 1961. The diversity of pollination systems in the Phlox family. *Recent Advances in Botany* (*Toronto*) **1**: 55–60.

GRANT, V., and A. GRANT. 1954. Genetic and taxonomic studies in Gilia. VII. The Woodland Gilias. *Aliso* **3**: 59–91.

GRAY, A. 1870. Revision of the North American Polemoniaceae. *Proc. Amer. Acad. Arts. Sci.* **8**: 247–282.

HEGI, G. 1927. Polemoniaceae. In his *Illustrierte Flora von Mittel-Europa* **5** (3): 2111–2119. Lehmanns Verlag, Munich.

HOOKER, J. D. 1869. *Cobaea penduliflora*. *Bot. Mag.* **25**: t. 5757.

HURD, P. D., and C. D. MICHENER. 1955. The megachiline bees of California (Hymenoptera: Megachilidae). *Bull. Calif. Insect Survey*, 3.

JAEGER, P. 1954. Les aspects actuels du problèm de la chéiroptèrogamie. *Bull. Inst. Franc. Afrique Noire* (*Dakar*) **16**: 796–821.

KERNER, A. 1895. *The Natural History of Plants*. Vol. 2. Blackie, London. (Original German edition 1891.)

KNUTH, P. 1909. *Handbook of Flower Pollination*. Vol. 3. Oxford, London. (Original German edition 1899.)

KNUTH, P., and E. LOEW. 1905. *Handbuch der Blütenbiologie*. Vol. 3, Part 2. Engelmann Verlag, Leipzig.

KUGLER, H. 1963. UV-Musterungen auf Blüten und ihr Zustandekommen. *Planta* **59**: 296–329.

LEPPIK, E. E. 1964. Floral evolution in the Ranunculaceae. *Iowa State Jour. Sci.* **39**: 1–101.

LEVIN, D. A. 1963. Natural hybridization between *Phlox maculata* and *Phlox glaberrima* and its evolutionary significance. *Amer. Jour. Bot.* **50**: 714–720.

LUDWIG, F. 1877. Ueber die Kleistogamie von *Collomia grandiflora* Dougl. *Bot. Zeitung* **35**: 777–780.

McDunnough, J. 1938. Check list of the Lepidoptera of Canada and the United States of America. Part 1. Macrolepidoptera. *Memoirs Southern Calif. Acad. Sci.*, 1.

Merritt, A. J. 1897. Notes on the pollination of some California mountain flowers, IV. *Erythea* 5: 15–22.

Muesebeck, C. F. W., K. V. Krombein, and H. K. Townes. 1951. *Hymenoptera of America north of Mexico. Synoptic catalog.* U.S. Dept. Agriculture, Monograph 2. Washington, D.C.

Müller, H. 1881. *Alpenblumen, ihre Befruchtung durch Insekten und ihre Anpassungen an dieselben.* Engelmann Verlag, Leipzig.

—— 1883. *The Fertilisation of Flowers.* Macmillan, London. (Original German edition 1873.)

Pennell, F. W. 1935. The Scrophulariaceae of eastern temperate North America. *Acad. Nat. Sci. Philadelphia, Monographs*, 1.

Peter, A. 1891. Polemoniaceae. In A. Engler and K. Prantl, *Die nätürlichen Pflanzenfamilien*, 4 (3a): 40–54. Engelmann Verlag, Leipzig.

Pigott, C. D. 1958. Biological flora of the British Isles. *Polemonium caeruleum* L. *Jour. Ecology* 46: 507–525.

Pijl, L. van der. 1936. Fledermäuse und Blumen. *Flora* 131: 1–40.

—— 1956. Remarks on pollination by bats in the genera Freycinetia, Duabanga and Haplophragma, and on chiropterophily in general. *Acta Botanica Neerlandica* 5: 135–144.

—— 1958. Flowers free from the environment? *Blumea*, Suppl. 4, 32–38.

—— 1960. Ecological aspects of flower evolution. I. Phyletic evolution. *Evolution* 14: 403–416.

—— 1961. Ecological aspects of flower evolution. II. Zoophilous flower classes. *Evolution* 15: 44–59.

Porsch, O. 1934–1935. Säugetiere als Blumenausbeuter und die Frage der Säugetierblume. I, II. *Biologia Generalis* 10: 657–685; 11: 171–188.

—— 1939. Das Bestäubungsleben der Kakteenblüte. *Cactaceae, Jahr. Deutsch. Kakteenges.* 1937: 1–142.

Ritzerow, H. 1907. Über Bau und Befruchtung kleistogamer Blüten. *Flora* 98: 163–212.

Robertson, C. 1891. Flowers and insects, Asclepiadaceae to Scrophulariaceae. *Trans. Acad. Sci. St. Louis* 5: 569–598.

—— 1895. Flowers and insects. XIV. *Bot. Gaz.* 20: 139–149.

—— 1928. *Flowers and Insects.* Lists of Visitors of 453 Flowers. Privately printed, Carlinville, Illinois.

Ross, H. 1898. Blüthenbiologische Beobachtungen an *Cobaea macrostemma* Pav. *Flora* 85: 125–134.

Scharlok, T. 1878. Ueber die Blüthen der Collomien. *Bot. Zeitung* 36: 640–645.

Scholtz, M. 1893. Die Orientirungsbewegungen des Blüthenstieles von *Cobaea scandens* Cav. und die Blütheneinrichtung dieser Art. *Cohn's Beiträge zur Biol. der Pflanzen* 6: 305–336.

Sprengel, C. K. 1793. *Das entdeckte Geheimniss der Natur im Bau und in der Befruchtung der Blumen.* Berlin.

Standley, P. C. 1924. Trees and shrubs of Mexico. *Contrib. U.S. Nat. Herb.*, 23.

Stebbins, G. L. 1950. *Variation and Evolution in Plants.* Columbia Univ. Press, New York.

—— 1957. Regularities of transformation in the flower (review). *Evolution* 11: 106–108.

TAKHTAJAN, A. 1959. *Die Evolution der Angiospermen.* Gustav Fischer Verlag, Jena.

TIMBERLAKE, P. H. 1954. A revisional study of the bees of the genus Perdita F. Smith with special reference to the fauna of the Pacific coast. *Univ. Calif. Publ. Entomol.* **9:** 345–432.

VOGEL, S. 1954. Blütenbiologische Typen als Elemente der Sippengliederung dargestellt anhand der Flora Südafrikas. *Botanische Studien,* 1. Gustav Fischer Verlag, Jena.

—— 1958. Fledermausblumen in Südamerika. *Österr. Bot. Zeitschr.* **104:** 491–530.

—— 1959. Organographie der Blüten kapländischer Ophrydeen mit Bemerkungen zum Koadaptions-Problem. *Akad. Wissensch. u. Lit. Mainz,* 1.

WERTH, E. 1956. *Bau und Leben der Blumen.* Enke Verlag, Stuttgart.

WHERRY, E. T. 1931. The eastern long-styled Phloxes. I. *Bartonia* **13:** 18–37.

—— 1932. The eastern long-styled Phloxes. II. *Bartonia* **14:** 14–26.

—— 1933. The eastern veiny-leaved Phloxes. *Bartonia* **15:** 14–26.

—— 1936. Miscellaneous eastern Polemoniaceae. *Bartonia* **18:** 52–59.

—— 1944. Review of the genera Collomia and Gymnosteris. *Amer. Midl. Nat.* **31:** 216–231.

—— 1946. The *Gilia aggregata* group. *Bull. Torrey Bot. Club* **73:** 194–202.

—— 1955. *The Genus Phlox.* Morris Arboretum Monographs, Philadelphia.

—— 1961. Remarks on the *Ipomopsis aggregata* group. *Aliso* **5:** 5–8.

INDEX

Page numbers of illustrations of plant and animal species are in italics. For references to particular insects, birds and bats, consult first the Systematic List of Animal Visitors (pp. 167 ff.), then look up the corresponding plant species in this index.